Electronic Experiences in a Virtual Lab

Roberto Gastaldi · Giovanni Campardo

Electronic Experiences in a Virtual Lab

 Springer

Roberto Gastaldi
Milan, Italy

Giovanni Campardo
Bergamo, Italy

ISBN 978-3-030-45181-3   ISBN 978-3-030-45179-0 (eBook)
https://doi.org/10.1007/978-3-030-45179-0

This Springer imprint is published by the registered company Springer Nature Switzerland AG
The registered company address is: Gewerbestrasse 11, 6330 Cham, Switzerland

Foreword

Roberto and Giovanni, the authors, contacted me about their idea to write this book on analog design, which has a very practical approach. This book should enable the reader to experiment with and experience the theoretical topics. This aligns with the aim of the organization I lead in EMEA, the Cadence Academic Network. Our goal is to 'lower the barriers to accessing leading-edge technology' for the next generation of engineers. Therefore, the Cadence Academic Network has joined forces with the authors in order to provide practical experience using Cadence's OrCAD® software to the reader of this book.

Part of the full OrCAD® package is the Cadence® PSpice® Simulator for analog and mixed-signal designs as well as OrCAD® Capture. Using these tools, the reader will be able to accomplish all practical laboratories, which are provided at the end of each chapter. The reader is able to download the fully functional trial version of the OrCAD® package here: https://www.orcad.com/free-trial. The trial version expires after 14 days, with a possible extension. Academic readers, who have a valid university email address and matriculation certificate, can get one-year access to the full version of the OrCAD® package through the OrCAD Academic Program: https://www.orcad.com/orcad-academic-program. For usage of the software, a PC with 8 GB RAM, 50 GB free hard drive space and 64 bit Microsoft® Windows®, version 7 or later, is required.

I hope you have a great experience reading this book and executing the experiments using OrCAD® software! For further information on the OrCAD® software, please visit https://www.orcad.com where you can find multiple tutorials, user forums, blogs, FAQs and other useful resources.

Anton Klotz
University Program Manager EMEA
Cadence Design System
Feldkirehen, Germany
aklotz@cadence.com

Preface

I was a teacher in a university for a couple of years in a basic electronic study course. I liked this experience, and the didactical approach was really interesting and exciting. The classes were attended by 25 people, all beginners in electronics, and the classroom space was big with big laboratory tables, a table each with 2 or 3 students, maximum. Each table was equipped with some instruments: an oscilloscope, a waveforms generator, a power supply, a soldering tool and wires. Electronic components like resistors, capacitors and integrated circuits were available in some cabinets in the laboratory. The duration of the lesson was around 4 hours. It seems a lot, but it is not so true; the session was scheduled in two parts. Every session was divided into three different parts. During the first part, I draw a circuit on the blackboard, discussing how the different parts work and all the questions related to them. I also sketched the equations to show the relationship between different parameters and circuital connections and component values. After that, the students started to build or design, depending on the previous theoretical explanation, their own circuits, using a breadboard to assemble or soldering the components, and they try to put in place what we explained in the theory. Last step was to show to all the class the results of the job.

In my opinion, this is the best way to proceed, as we can say: '…if I listen, I forget, if I see, I remember, if I do, I understand…'.

One of the most important things to understand is that we live with a lack of exactness and with the conflict of theory versus practice. I remember a student who wrote tons of paper of equations to find the 'right' value for the components to bias a transistor, forgetting that the components are available only with the standard values and their precision or tolerance, especially in a student laboratory, for a problem of cost, is always around 20%. So, after an hour of calculation, he discovered that, to have the right bias, we need to change the value of many components or add more of them. This is the difference between theory and practice.

Also, to discover that if you connect an electrolytic capacitor on the opposite polarity, it will burn and can also explode reaching the roof of the laboratory. A big hole in the roof shows to all the students, forever, what can happen if we suspend our attention.

Today, we can propose to replicate the lessons using a virtual laboratory, an important software that replicates the instrumentation needed to build and test a circuit.

This is a collection of a number of 'lessons' about some topics commonly encountered in electronic circuit design.

Actually, they are not lessons in the usual meaning given to this word because generally speaking, a lesson starts from the explanation of general principles on which the design of a circuit is based, and then, a particular circuit realization is carried out making use of those principles. On the contrary, we would like to reverse this approach and start from a function realized with a complex electronic circuit as used in the common engineering practice and take that as a vehicle to explain the basic circuits from which it is made of and explain physical laws which drive the operation of those circuits. For this reason, the reader must have basic skills in electronics, acquired in basic courses.

By experience, the first step is to study the fundamental blocks, components and general laws. To design this, knowledge is not enough. The second step is to study circuits, already designed by previous engineers to solve different problems. The last step will be to design new circuits. Our book is devoted to the second step. To do this, again by our experience, the laboratory practical is a fundamental approach. Today, thanks to the old engineers, we can have a 'virtual laboratory on our desk', a circuital simulator. For this reason, we had a contact with OrCAD, design SW company, to have the possibility to use their SW simulator and give to the readers, acquiring the book, to use OrCAD tools to reproduce the same virtual experiments we will do in the book. The steps of this process can be summarized in this way:

1. The complex circuit under examination is captured on OrCAD environment. The resulting network is shown in a figure.
2. Basic circuits forming the complex example under evaluation are detected, and their function for the operation of the whole system is explained.
3. Stimuli are added to the network, and the simulation tool (PSpice) is launched. Resulting waveforms of critical points of the network are shown in some figures
4. Value of the most critical parameters is changed and is simulated again. Waveforms of critical nodes are compared with previous ones and changes are discussed.
5. Principles of circuit theory are used to explain the results obtained, and the sensitivity of critical parameters and a general rule for the operation of the circuit under examination are found.
6. New analysis is suggested to the reader to learn more about the topic.

We selected a number of case studies to which this methodology is applied. We chose them among a great number of 'basic' circuit configurations so that they can give a survey on enough number of important building blocks in electronic design practice.

Some comments:

Chapter 1 is a quick summary of the basic principles of electronics commonly used in circuit design practice.

Chapter 2 is devoted to the analysis of a couple of power supplies. We chose to start from them because in spite of their importance in the common applications, they are often neglected in design manuals.

Chapter 3 is dedicated to the amplifiers. Of course, amplification of a signal is a fundamental function of almost all of the electronic circuits. In this chapter, first a simpler case of a multistage voltage amplifier using discrete transistor devices is analyzed. A second example is devoted to the study of a feedback amplifier, understanding through simulation how this technique can greatly improve amplifier characteristics. The third example is a linear amplifier using a differential pair as input stage. This configuration is very important because it is the basis of the operational amplifier which is the building block of all the analog electronic circuits on the market.

Chapter 4 enters into more details about the application of analog electronics through the analysis of a more complex circuit which performs sound equalization like it is used in a hi-fi sound amplifier. It is also an opportunity to talk about processing (filtering) an analog signal.

Chapter 5 introduces logic circuits through the analysis of counters first and then the implementation of an electronic clock. Finally an example will be given of an adder which is a key component of a computer CPU.

Chapter 6 is dedicated to oscillators which are circuits capable of generating a waveform, both sinusoidal or square or other. They are not only the most important components of radio transmitters but also the pulsing heart of our computer (the famous clock which sets the computer speed). One of the examples of this book is just the design of a clock circuit.

Chapter 7 is a collection of three examples of complex circuits which are analyzed starting from their basic sub-blocks, then putting them together to obtain the final scheme. They are a good example of how a complex electronic circuit is designed and simulated.

Chapter 8 deals with the simulation of the characteristics of an active device like a diode or a transistor. Normally, this is not required as they are already included in the component library of the simulation tool, but sometimes it can be useful to have a graphical representation of these characteristics to better understand the behavior of the simulated circuits.

Chapter 9 is dedicated to explain how the OrCAD simulation tool works.

Before starting, we would like to underline that all the people working in a technology environment have, as the main target, the capability to make predictions. What does it mean?

Thanks to many special people, like Archimedes, Maxwell, Fermi, Gauss, Einstein, Newton and many others, we know the laws that govern the world and we use these laws to imagine many applications to improve our life.

For different reasons, we need to be able to do the calculation to predict if the airplane we are designing will flight or not. We cannot wait 'to try.' We need to be sure before to take off from the airport. To guarantee the success of our 'application,' we do, through calculation, ideal tests to simulate different real conditions.

Today, with the help of computers, we can simulate a lot of components, and using different tools we can perform electrical simulations or mechanical, thermal, hydrodynamic, or other.

What we need is a library of components where a model, for any different component, exists. Then using the laws, we already know we can drive the computer to perform a lot of calculations to provide a simulation really near to reality.

Without these approaches, we will come back in the last century or even in the previous one.

The first suggestion is to take a pen and a paper and, before starting to design or simulate, copy the circuit and try to understand, for each component, what we are doing and why. For the simulation, always remember that simulation must 'only' confirm your idea. You have to know before the simulation what the circuit is doing. Simulation needs only to confirm the ideas and to calculate the right 'numbers' you need for your design and finally remember that no simulations can substitute a real laboratory session.

Let's start.

Bergamo, Italy   Giovanni Campardo
Milan, Italy   Roberto Gastaldi

Acknowledgements

We would like to acknowledge a number of persons whose contribution has been crucial for writing this book.

First of all, the whole staff at Springer and in particular Christophe Baumann, to whom we presented at first our project, who encouraged and supported us in the early phase of manuscript writing, and Pierpaolo Riva who assisted us and gave us useful advices during the final composition of this book.

Anton Klotz of Cadence who worked to give us an easy access to OrCAD PSpice simulation tool and ensured we could have a ready assistance during installation of the software in our computers. We specially thank Anton to have written the foreword to this book.

Alok Tripathi of Cadence who accepted to write a chapter in this book dedicated to the use of PSpice Simulator and was a precious interface to quickly answer all our questions about simulator peculiarities.

We owe special thanks to Marco Brugora working at ST APG Automotive Digital Solutions who took the burden to review all the manuscripts. We thank a lot Marco for his deep and precise revision work. Thanks to his detailed remarks and advices, we were able to improve the manuscript as much as possible. We hope to have been successful in this task, but in any case, the responsibility for any mistake remains on us.

Finally, we would like to thank our wives who let us spend quite a lot of our free time in the last year to write this book.

Agrate Brianza, Italy   Giovanni Campardo
February 2020   Roberto Gastaldi

Contents

About the Authors

Roberto Gastaldi holds a Master degree in Electronic Engineering from Politecnico di Milano (Milan, Italy). He has been working for over thirty years in a semiconductor manufacturing corporation as a non-volatile memory designer. During his career he led the design of many successful memory chips as well as first prototypes for leading edge memory technologies. Mr. Gastaldi holds several papers and patents related to VLSI memory design and technology, he is a member of IEEE.

Giovanni Campardo was born in Bergamo, Italy, in 1958. He received the laurea degree in Nuclear Engineering from the Politecnico of Milan in 1984. In 1997 he graduated in Physics from the Università Statale di Milano. He deals with micro-electronics, today in an Italian-French company, in particular in the field of non-volatile solid state memories. He is author-co-author of many publications and some books and hold more than 100 patents. He is married with two children and enjoys painting and reading. He collects ethnic masks from the entire world and practice Chinese martial arts.

Chapter 1
Review of Some Basic Concepts

1.1 Introduction

The characteristic of this book is to teach some topics of electronic design through the analysis of real circuits understanding their behavior from simulation results as if you really had built your circuit in a laboratory and measured it with a scope. Simulation is usually user-friendly; you haven't to physically change components and connections to make a modification, so it is easy to start to make a lot of trials more or less without having really understood how the circuit works. This methodology leads to a terrible waste of time and doesn't bring the best results. Instead, reviewing the results applying the theory of electronic design leads to focused trials which give the best results in the shortest time. Theory of electrical networks and circuit design is incredibly wide and even a dedicated book should not be enough to give a complete overview of such a topic. Of course, the scope of this short chapter is by far more limited, the idea is simply to recall some of the basic electronic theory principles, so you can be conscious of the results you get from simulation and of the reason why a certain change produces a certain effect. Then, let's start!

1.2 Resistor

We all know from physics that current flow through a material, for example, a metal wire, is characterized from a voltage drop between two points A and B of the wire. Ohm's law states that the ratio between the voltage drop from A to B and the current (I) flowing in the wire is called resistance (R).

$$R = \frac{V_{a,b}}{I} \qquad (1.1)$$

© Springer Nature Switzerland AG 2020
R. Gastaldi and G. Campardo, *Electronic Experiences in a Virtual Lab*,
https://doi.org/10.1007/978-3-030-45179-0_1

Resistance is a quantity which depends on the geometry and on the characteristics of the material we are considering. For example, if we have a wire whose length is L and with a section S, its resistance is:

$$R = \frac{\rho L}{S} \tag{1.2}$$

Greek letter ρ stands for 'resistivity' a quantity dependent on the particular material used. Equation (1.2) tells us that the resistance is directly proportional to the length of our wire and inversely proportional to his section.

While a wire used to connect the different parts of an electronic circuit is made of a low-resistivity material (copper for example), sometimes in the electronic circuits it is necessary to introduce intentionally a resistive element. Most common resistors are made with a mixed carbon closed in a cylindrical box or with a ceramic case on which a carbon layer is deposited. Generally, the resistance real value of these components has a tolerance with respect to his nominal value. Most common tolerance is 5 or 10%. Tolerances of 1–0.1% are also possible but require a different fabrication technique.

1.3 Capacitor

A capacitor is conceptually composed of two conductive plates with a dielectric material interposed. When a voltage source is connected across the two plates, a positive charge is accumulated on a plate and an equal negative charge on the other, so that a potential can be measured across the plates. The most important relationship between voltage across a capacitor and current flowing inside it is the following:

$$I(t) = C \frac{\mathrm{d}V}{\mathrm{d}t} \tag{1.3}$$

This simple relationship tells us that the current flowing into a capacitor is directly proportional to the rate of variation of the voltage across it, and the proportionality constant is the capacitance C. We also can say that if the voltage across the capacitor is stable (no variation with time), the current flow is zero. To empirically understand the behavior of a capacitor in a network subject to time-varying signals, you can keep in mind that *the voltage across a capacitor cannot change instantaneously*.

If $V(t)$ can be written as $V_0 \sin(t)$, the current will be $CV_0 \cos(t)$, and if we plot this waveform, we can recognize that there is a 90° delay between current and voltage.

We can clearly understand this point building in our virtual laboratory the very simple circuit of Fig. 1.1 and simulating with our PSpice tool.

The resulting waveforms in Fig. 1.2, neglecting the two different scales, clearly show that the voltage across the capacitor has a delay with respect to the current flowing through it. In terms of angular coordinates, this delay is 90°.

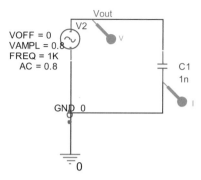

Fig. 1.1 Capacitor connected to a sinusoidal voltage generator

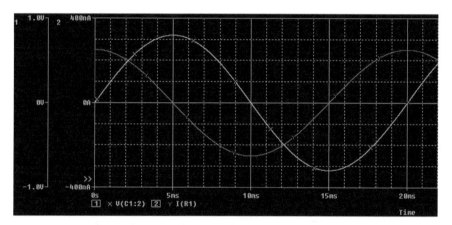

Fig. 1.2 Sinusoidal voltage across a capacitor (green) which has a 90° delay with respect to the current (red)

The capacitance is measured in Farad (F), but in electronic circuits, very small capacitors are used in the order of 10^{-12} F up to 10^{-6} F.

1.4 Inductor

A number n of coils around a material with high magnetic permeability μ generates a magnetic field H when a current $i(t)$ flows through them (see Fig. 1.3).
Physical theory tells that:

$$H = \frac{B}{\mu} \tag{1.4}$$

Fig. 1.3 Schematization of
an inductance

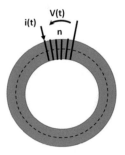

where B is the magnetic induction. Considering the circle inside the material of length
l (see the dotted line in Fig. 1.3), the theory tells that:

$$Hl = ni \tag{1.5}$$

This easily becomes:

$$Bl = n\mu i \tag{1.6}$$

We assume that the magnetic field is completely included and uniform through
the area A of a single coil, corresponding to the section of the toroidal material, and
then, the magnetic flux ϕ concatenated with the n coils is:

$$\phi = n\mu i \frac{A}{l} \tag{1.7}$$

When the current through the coils changes the magnetic flux associated with
a single coil changes, then, due to electromagnetic induction on each coil, an
electromagnetic force $e(t)$ is generated in such a way to counteract the current
variation:

$$e(t) = n^2\mu \frac{A}{l} \frac{di}{dt} \tag{1.8}$$

Now if we define the inductance L as:

$$L = n^2\mu \frac{A}{l} \tag{1.9}$$

*e(t) makes the voltage at the terminal of incoming current to rise to limit the
current increase so that*:

$$v(t) = L\frac{di}{dt} \tag{1.10}$$

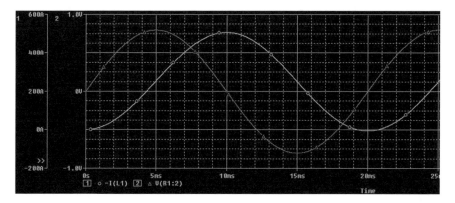

Fig. 1.4 An inductor exercised with a sinusoidal waveform. This time the current is in delay with respect to the voltage

We also see from this relationship that the faster is the variation of the current through the inductance L the higher is the voltage across the coils. The behavior of inductance is dual with respect to what we have seen for the capacitor, so this time *the current flowing into an inductance cannot change abruptly*. If the current flowing in the circuit is suddenly stopped at time $t = 0$, a voltage rises across the inductance to maintain the current unchanged at time $t = 0^+$.

Also for inductance, the relationship between current and voltage becomes very simple considering the case of a sinusoidal current: In this case, the current across the inductance is still sinusoidal but $90°$ in advance with respect to the voltage. Using our simulator, it is easy to verify this statement, just replacing the capacitor with an inductor in the circuit of Fig. 1.1. The result is shown in Fig. 1.4.

1.5 Kirchhoff's Laws

They are the fundamental laws to study all the electrical networks:

(I)   The algebraic sum of all the currents flowing into wires crossing in a node is zero.
(II) The algebraic sum of the voltages measured between the nodes of a net is zero.

Using (I) and (II) combined with Ohm's law (1.1), we are able to write linear systems of equations to solve complex electrical networks when capacitors and inductors are not involved or their effect can be neglected. Otherwise, we need to introduce integral/differential equations to solve the network.

It is important to underline that these laws are verified only if the network is in a steady-state condition or the electrical signal changes 'slowly' so that capacitive and inductive interactions between the conductors can be reasonably neglected.

Fig. 1.5 Simple resistive
network: $R_1 = 1\ \Omega, R_2 = 1\ \Omega, R_3 = 3\ \Omega$

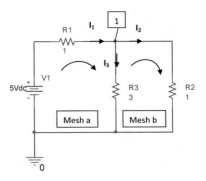

Let's see a small example of how Kirchhoff's laws and Ohm's law are used to solve electrical networks considering the simple network of Fig. 1.5.

This is a resistive network operated in DC mode with the voltage generator V_1. We want to calculate the currents I_1, I_2, I_3 circulating in the network and the voltages across each resistor of the network using Kirchhoff's laws and Ohm's law. We start observing that there are two meshes and two nodes in the network (forget the node connected to ground that is necessary for simulation purposes but is not really needed for the network). We write two equations for meshes and one for one of the two nodes, resulting in a system of three linear equations in three unknowns. If the equations are independent, we are able to find the three unknowns. Let's write them having in mind that the voltages of the meshes are taken as positive moving clockwise (see the arrows in Fig. 1.5), and the currents are taken as positive when they enter the node.

Kirchhoff's law at mesh a): $I_1 1 + I_3 3 = 5$
Kirchhoff's law at mesh (b): $I_2 1 - I_3 3 = 0$
Kirchhoff's law at node 1: $I_1 1 - I_2 1 - I_3 1 = 0$

We can write these relationships in such a way that the three unknowns appear in all the equations, like this:

$$\begin{cases} I_1 1 + I_2 0 + I_3 3 = 5 \\ I_1 0 + I_2 1 - I_3 3 = 0 \\ I_1 1 - I_2 1 - I_3 1 = 0 \end{cases}$$

This is a matricial form of our system that can be written in a compact mode as:

$$\boldsymbol{AI} = \boldsymbol{b} \tag{1.11}$$

where

$$A = \begin{bmatrix} 1 & 0 & 3 \\ 0 & 1 & -3 \\ 1 & -1 & -1 \end{bmatrix}$$

$$b = \begin{bmatrix} 5 \\ 0 \\ 0 \end{bmatrix} \quad I = \begin{bmatrix} I_1 \\ I_2 \\ I_3 \end{bmatrix}$$

We can find the determinant of 3×3 matrix A applying the rules of matrix algebra finding:

$$Det\ A = -7$$

Mathematical theory tells that the solution of this system is given by the formulas:

$$I_1 = \frac{Det \begin{bmatrix} 5 & 0 & 3 \\ 0 & 1 & -3 \\ 0 & -1 & -1 \end{bmatrix}}{Det\ A}$$

$$I_2 = \frac{Det \begin{bmatrix} 1 & 5 & 3 \\ 0 & 0 & -3 \\ 1 & 0 & -1 \end{bmatrix}}{det\ A}$$

$$I_3 = \frac{Det \begin{bmatrix} 1 & 0 & 5 \\ 0 & 1 & 0 \\ 1 & -1 & 0 \end{bmatrix}}{Det\ A}$$

Completing the calculations, we come at the end to the values of the currents:

$$I_1 = \frac{20}{7} = 2.857\ \text{A}; \quad I_2 = \frac{15}{7} = 2.143\ \text{A}; \quad I_3 = \frac{5}{7} = 0.7143\ \text{A}$$

Now it is very simple to obtain the voltages across the resistors of the network simply applying Ohm's law:

$$V_{R1} = 2.857 \times 1 = 2.857\ \text{V}; \quad V_{R2} = 2.143 \times 1 = 2.143\ \text{V}; \quad V_{R3} = 0.7143 \times 3$$
$$= 2.143\ \text{V}$$

Well, we finally have solved our network! Shall we compare our results with the output of simulation? Please have a look to Fig. 1.6

The correspondence is perfect. (Remember that the voltage across R_1 is given by $5.0 - 2.143 = 2.857$ V.)

Nevertheless if we think about the procedure we have followed we understand that it is quite cumbersome and requires a lot of calculations even for a very simple network as the one we considered. You can imagine what happens if we consider a

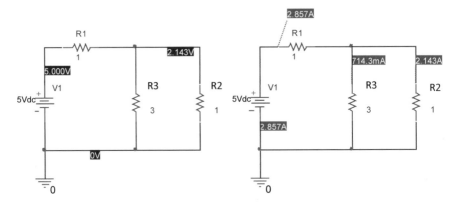

Fig. 1.6 Voltages (left side) and currents (right side) at the nodes of the network found using the simulator

more complex network! If we want to solve the problem using the human brain, it is mandatory to try some way to simplify the network so that its complexity and the amount of calculations can be substantially reduced.

Question: *can you find a way to really simplify the calculations on the network we considered*?

If the elements like inductors and capacitors are included in the network, the problem of solving is even more difficult because integral–differential equations are involved. Luckily, we can use a technique called *Laplace transform* which allows downgrading differential equations to linear equations. We will not develop this topic in these short notes; otherwise, we should have to dedicate a whole book for it!

On the other hand, if you use a computer to find currents and voltages of a network, solving a linear system of equations in a matricial form is much easier than to find network simplification, as the computer can do calculations at very high speed, provided that it can deal with simple, repetitive operations.

1.6 RC Network

A very common circuit we can see working with electrical networks is shown in Fig. 1.7. The simple network is stimulated by a step voltage V_1, in other words:

$$\begin{Bmatrix} V_1 = 0 & \text{for} & t \leq 0 \\ V_1 = 5 & \text{for} & t > 0 \end{Bmatrix}$$

It is interesting to analyze the behavior of the voltage across the capacitor as a response to a step voltage applied with the input generator. We can write the equation related to the only mesh of the circuit:

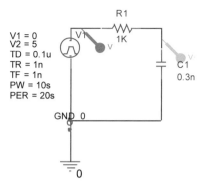

Fig. 1.7 Simple RC network

$$V_1 = R_1 C_1 \frac{dV_{C1}(t)}{dt} + V_{C_1} \tag{1.12}$$

Solving this differential equation, we realize that the capacitor voltage cannot rise immediately but *gradually* with the following law:

$$V_{C1}(t) = V_1\left(1 - e^{-t/R_1C_1}\right) \tag{1.13}$$

This means that the step voltage applied at the input is not exactly reproduced at the output (see Fig. 1.8). To obtain a more exact replica of the input waveform, it is necessary to reduce the quantity R_1C_1 that we define *time constant* τ.

$$\tau = R_1 C_1 \tag{1.14}$$

The circuit of Fig. 1.5 is often called an *RC delay*.

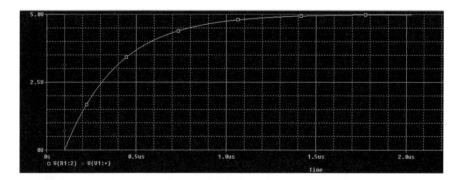

Fig. 1.8 Step voltage (red) applied to a RC circuit results into a smooth waveform (green) at the output

We can run a simulation of the circuit, and we obtain the waveforms shown in Fig. 1.8 for the step voltage and the voltage across the capacitor. We could verify that it fits Eq. (1.13)

Question: *The voltage generator used in* Fig. 1.8 *to approximate a step voltage is actually a square wave generator with a period much longer than RC. Try to reduce the period to a value comparable with RC and look at the results. Can you write the mathematical expression of the output (green curve) when the input makes a transition from 5 V to zero?*

The time constant τ is linked to the cut frequency of an electrical network f_C when a sinusoidal signal is applied from the relationship:

$$\tau = \frac{1}{2\pi f_C} \tag{1.15}$$

For example, the network of Fig. 1.7 has a f_C of 530.7 kHz calculated with the above formula. A practical way to understand what it means is to apply to the input a sinusoidal signal with increasing frequency and observe the voltage across the capacitor (V_{out}) which is still a sine wave but with amplitude as in Fig. 1.9

You can see that the voltage across the capacitor is constant and equal to the input source at low frequency, but from about 100 kHz on starts to decrease. The vertical red line marks $f_C = 530.7$ kHz, and at that point, the voltage amplitude is $V_{out} = 0.707\ V_{in}$. This suggests a definition for f_C: It is the frequency at which a sinusoidal wave amplitude is reduced of a factor 0.707 (or $1/\sqrt{2}$) passing through the network. From Fig. 1.9, it can say that our RC network is a low-pass filter that attenuates all the frequencies higher than f_C.

It is interesting to substitute an inductor in place of the capacitor C_1 of the circuit in Fig. 1.7. The key concept to understand the operation of this new network is that the current flowing into the inductor cannot change abruptly, then assuming that initially $i_L = 0$ the sharp variation of V_1 $\left(\left\{ \begin{array}{l} V_1 = 0\ t \le 0 \\ V_1 = 5\ t > 0 \end{array} \right\} \right)$ will be transmitted unaltered

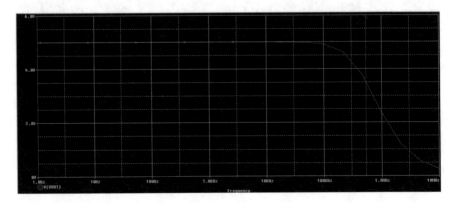

Fig. 1.9 Effect of frequency cut on a RC network

at the terminals of the inductor: In other words, the impedance of the inductor at $t = 0^+$ is ∞. However gradually, the current starts increasing into it, so that at the end, the voltage across it will be zero. This circuit can be seen as a high-pass filter which blocks low frequencies.

Questions: Can you write the mathematical expression of the current flowing in the mesh and the voltage across the inductor? The problem is similar to the one discussed above with capacitor.

Repeat the frequency sweep of Fig. 1.9 on the inductor circuit and discuss the result.

1.7 Resonant Networks

Let's consider a network like in Fig. 1.10.

It is made of a resistor, an inductance and a capacitor connected in parallel and excited with a sinusoidal current generator of 10 kHz frequency (I_1). If we calculate the admittance $Y(\omega)$ of this circuit, we can write:

$$Y(\omega) = \frac{1}{R_1} + j\left(\omega C_1 - \frac{1}{\omega L_1}\right) \tag{1.16}$$

In the particular condition in which $\left(\omega C_1 - \frac{1}{\omega L_1}\right) = 0$, or in other words when $\omega = \omega_0 = \frac{1}{\sqrt{L_1 C_1}}$ the admittance has a minimum, then the voltage across $R_1 L_1 C_1$ has a sharp maximum as shown in Fig. 1.11

Fig. 1.10 Schematic of a parallel RLC network

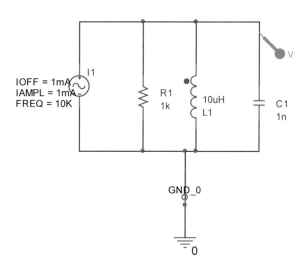

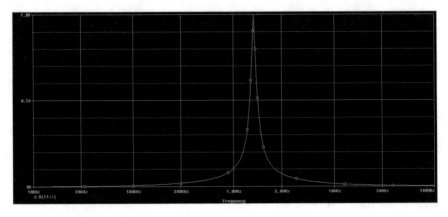

Fig. 1.11 Voltage across RLC circuit during a frequency sweep

Actually $f_0 = \frac{\omega_0}{2\pi}$ is called *resonance frequency*, and at that frequency, the impedance of the circuit is purely resistive and equal to R_1 which is in parallel with infinite impedance.

In the practical case, we that are considering, $f_0 = 1.59$ MHz and the voltage peak should be the product of 1 Kohm multiplied by 1 mA, then 1 V. You can see that the simulation results shown in Fig. 1.9 are well in accordance with these calculations.

A dual configuration of the circuit of Fig. 1.10 is shown in Fig. 1.12. Here, the RLC components are connected in series and a voltage generator is connected instead of a current one. In this case, it is the impedance $Z(\omega)$ that is minimized at f_0 and we can detect a peak in the current flowing into the circuit.

Exercise: *Try to simulate the circuit of* Fig. 1.12 *and plot the behavior of the current inside the circuit as the frequency changes.*

Fig. 1.12 Series resonant RLC

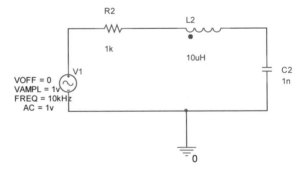

1.8 Operation Amplifier

Operational amplifiers (Op Amp from now on) are one of the most diffused components in analog electronic systems and the analysis of operation of basic circuits using this component can be very useful also to recall important principles of analog electronics, and then this section is devoted to quickly review operational amplifier applications.

An Op Amp is a gain stage with differential input that ideally has the following characteristics:

- Input impedance $= \infty$
- Output impedance $= 0$
- Voltage gain $= -\infty$
- Bandwidth $= \infty$

Let's see an example of use of an Op Amp to build a voltage amplifier: We can draw a schematic like in Fig. 1.13.

If we use our simulator, we will have to use modeled real components, but for now we can assume the above condition of ideality is met. In this case, we can start to observe that pin 2 of the Op Amp is about ground because voltage gain of the Op Amp is $-\infty$ and feedback resistor R_2 forces $V_2 - V_3 = 0$. Then, we can write:

$$I_{R1} = \frac{V_3}{R_1} \tag{1.17}$$

On the other hand, ideality hypothesis requires that no current can enter the Op Amp through terminal 2, so that all the current is flowing through the resistor R_2, following the hypothesis that terminal 2 is at ground potential the voltage across R_2 is $-V_{\text{out}}$ where the sign takes into account that the amplifier is inverting the input signal, then:

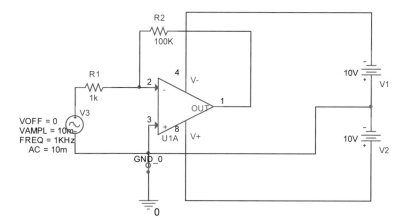

Fig. 1.13 Classical inverting voltage amplifier made using an operational amplifier

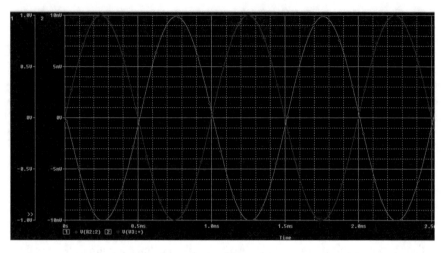

Fig. 1.14 Simulation of the inverting voltage amplifier of Fig. 1.13. The input is in blue and the output in red (see the probes in Fig. 1.13)

$$\frac{V_3}{R_1} R_2 = -V_{\text{out}} \tag{1.18}$$

Or if we call $V_3 = V_{\text{in}}$

$$\frac{V_{\text{out}}}{V_{\text{in}}} = -\frac{R_2}{R_1} \tag{1.19}$$

Simulating the circuit of Fig. 1.13, we will find the output shown in Fig. 1.14

In spite of the fact that we have used a real device in which voltage gain $\neq \infty$, the voltage gain of our circuit fits very well, and the relationship found above because even the gain of our real Op Amp is very high. Nevertheless let's try to build the exact equivalent circuit for this case. To make an example, we can consider an Op Amp with open-circuit voltage gain $A_v = -100$ K where the sign minus indicates that it is an inverting amplifier, let's suppose again that the output resistance $R_0 = 100\ \Omega$. We can draw a model as in Fig. 1.15

In this figure, you can recognize that the Op Amp has been modeled as a voltage controlled-voltage generator whose gain is 100 K. The inverting behavior of the amplifier has been modeled connecting the controlling voltage to the negative side of the controlling terminals E_1. All the other elements are as in Fig. 1.11 except that resistance R_2 was replaced with two resistances R_a and R_b, and the output resistance of the Op Amp (R_0) *has* been added. We can notice that the introduction of R_a and R_b greatly simplify the solution of the network, and this has been possible because of a very important theorem called the *Miller theorem*. Thanks to it if we have a generic impedance Z_f connected between two points A and B and we know the voltage ratio $A_V = \frac{V_B}{V_A}$ we can replace that impedance with two others:

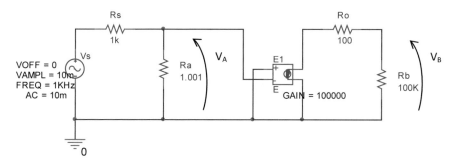

Fig. 1.15 Model of the inverting amplifier of Fig. 1.13

$$Z_a = \frac{Z_f}{(1 - A_V)} \tag{1.20}$$

$$Z_b = \frac{Z_f A_V}{(A_V - 1)} \tag{1.21}$$

In the case, we are considering A_V is the gain of our amplifier but keeping into account the output resistance, then:

$$A_V = A_v + \frac{R_0 Y_f}{\left(1 + R_0 Y_f\right)} \tag{1.22}$$

where $Y_f = \frac{1}{Z_f}$. Of course, in case $R_0 = 0$ then $A_V \equiv A_v$

Exercise: *Use the data we have considered to verify the values of R_a and R_b in Fig. 1.15 and to calculate A_V.*

Now we can calculate the gain of the amplifier stage model of Fig. 1.15, and in other words we want to know

$A_{\text{amp}} = \frac{V_0}{V_s}$. With some boring but not difficult calculations, we can demonstrate that it is given by:

$$A_{\text{amp}} = -\frac{Y_s}{\left(Y_f - \left(\frac{1}{A_V}\right)(Y_f + Y_s + Y_i)\right)} \tag{1.23}$$

where $Y_s = \frac{1}{R_s}$, $V_o \equiv V_B$ and Y_i stands for the input admittance of the Op Amp that we have considered as zero (which means infinite input resistance).

We can input the data in this formula, we have:

$Y_s = 0.001$ mho
$Y_f = 0.00001$ mho
$|A_V| = 99.9$ K
$Y_i = 0$

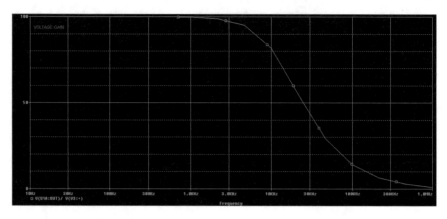

Fig. 1.16 Plot of voltage gain versus frequency of the circuit in Fig. 1.13

We can immediately see that with these values, the correction to the approximate calculation we made above is absolutely negligible. Instead, it is very important to remember Miller's theorem because it has a lot of applications in electronic design.

Now, instead of making all the calculations by hand we could have used our simulator. Let's try and see: the result is shown in Fig. 1.16.

Here, I have made a plot of voltage gain vs. frequency because we can see better the behavior of the circuit: At low frequency, the gain is 100 as predicted by our calculation (great!), but at high frequency, the gain starts to roll off and becomes about zero at 1 MHz. This is due to the fact that the real Op Amp has a finite bandwidth. The value of R_2 affects the bandwidth:

Reduce the value of R_2, in this way the gain is reduced. What happens to the bandwidth? Now try the opposite and increase R_2. Discuss the results.

What happens if you put a capacitor in parallel to R_2?

Done? Fine, we can go on to the last section.

1.9   Analog Integration with Operational Amplifier

Operational amplifiers have a great versatility and can be used to build a wide number of analog blocks. The concept of *virtual ground* is a powerful tool to analyze circuits with Op Amp even if in approximate mode. In this section, I would like to give an example of an application using Op Amp to show the versatility of this component. We will consider the matter to integrate an analog voltage:

The schematic of an analog integrator is visible in Fig. 1.17

In summary, we can say that it is the same circuit shown in Fig. 1.13 where a capacitor has substituted resistance R_2. I can use the concept of virtual ground to write the equations of this network and write:

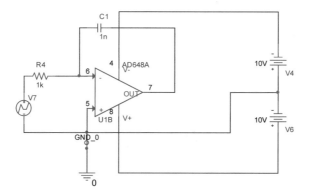

Fig. 1.17 Analog integrator

$$i_{R_4} = \frac{V_7}{R_4} \tag{1.24}$$

$$V_{OUT} = -\frac{1}{C}\int i_{R_4}dt = -\frac{1}{R_4 C}\int V_7 dt \tag{1.25}$$

In particular, if V_7 is a constant ($V_7 = V$), then the output will be a ramp:

$$V_{OUT} = -\frac{Vt}{RC} \tag{1.26}$$

Let's simulate the circuit of Fig. 1.15 in the case of constant input and look at the results in Fig. 1.18.

We can see that at a first sight, the output voltage is exactly a straight line as predicted with our approximate calculation, but probably things are a little different around the time $t = 20\ \mu S$. Please go through the proposed exercise here below to conclude this chapter.

Exercise: *Verify that the slope of output voltage is in accordance with what we found with calculations. Then focus at the beginning of the ramp. Enlarge the output waveform around that point and try to explain the behavior you find. Remember that the simulation has been done using a real component, what would be the behavior of an ideal Op Amp with infinite bandwidth?*

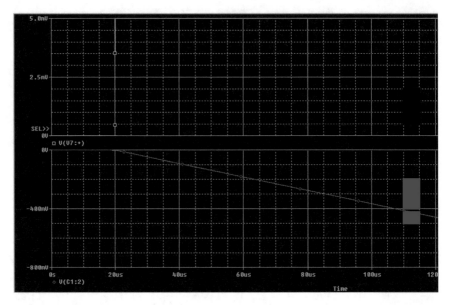

Fig. 1.18 Results of analog integrator circuit simulation: in green constant input voltage starting from $t = 20$ uS, in red the linear increasing output

Chapter 2
Power Supplies

This chapter is devoted to the analysis of the power supplies. We choose to start from these types of circuits, for this book, because, in spite of their importance, in the common applications, they are often neglected in design manuals.

2.1 Introduction

Electronic students, normally, want to study, analyze and build circuit like amplifiers or transmitters and receivers, and generally, none is dedicated to the supply voltages, thinking that this argument is not so cool! So in real life, we will find out that, having an amplifier and no power is perfectly useless, of course, but also not having a good supply voltage can be a very serious problem. A supply voltage must not only guarantee a stable voltage output but also to be able to provide the right current required by the load, in all the situations. It must be able to filter the possible disturbance coming from the domestic power supply network or from the ether, and it must be able to filter the noise that the application itself could inject.

At the end, the supply voltage circuit works like all the other circuits: Take a signal, in this case the power supply network, and transform it to serve a load. Before starting to analyze some of the most common power supply circuits, we would like to spend few words about one point that we consider it very important, the usage of the decoupling capacitors.

In all the circuits, you can find capacitors connected from the supply voltage to the ground. These components are normally neglected, during the circuit operation discussion, but they cannot be neglected during the real circuit-working mode. These elements are used for decoupling that means to separate different sections in the circuit, to isolate different zones or to bypass that means to have a low impedance path to shunt to ground transient energy. The capacitors selection also should be driven considering a lot of parameters, as the DC current in the circuit section, the ripple or noise tolerable level; a wrong selection can produce worst or unwanted performance with respect to expected behavior. Of course, each component has additional parasitic

© Springer Nature Switzerland AG 2020
R. Gastaldi and G. Campardo, *Electronic Experiences in a Virtual Lab*,
https://doi.org/10.1007/978-3-030-45179-0_2

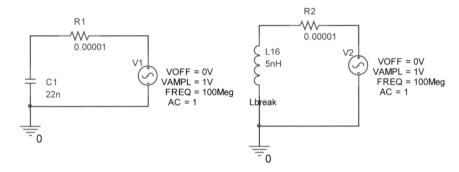

Fig. 2.1 Schematic of a capacitor and inductor to evaluate how the reactance varies with the frequency

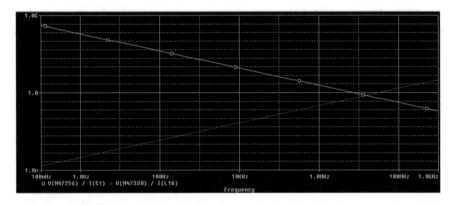

Fig. 2.2 X_C and X_L value variation with frequency

elements, due to connections, mounting, proximities to other components, etc., and we have to take these in account. This is the real world (Fig. 2.1).

Let me study a very simple configuration that all the readers of this book must know very well.

We would like to see how the reactance for the capacitor and for the impedance changes with the frequency. A very low resistor is added in series to permit the convergence of the simulator.

As we can see, Fig. 2.2, in the simulation, for an ideal capacitor, the impedance with frequency decrease tends to zero, green waveform, and for an ideal inductance, it tends to infinity, the red one. Brrr...none engineer can use zero or infinity as a value. This is not real.

Rules of thumb, coming from experience, says that for a component, like a capacitor, mounted on a printed circuit board, some additional effects and contributions must be considered: We have an inductance L contribution by 1 nH for each mm of wire, so for a 22 nF capacitor, we can imagine to have 5 nH of inductance and 30 mΩ

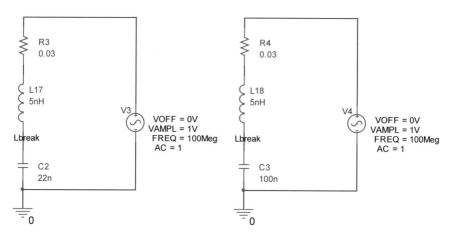

Fig. 2.3 Schematic of two different capacitors having the same parasitic values L and R

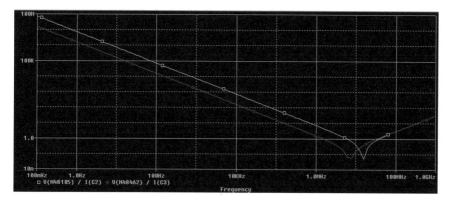

Fig. 2.4 Two capacitors, 22 nF the red one, and 100 nF, green, with the parasitic elements. Probe red and green are the impedance versus frequency of the capacitor plus its R and L parasitic

parasitic resistance, and of course, for a capacitor of 100 nF, the L and R parasitic will have same values as shown in Fig. 2.3. In the simulation, we have the behavior of the 'real' capacitor with the parasitic elements.

The real capacitor impedance has a different behavior, in frequency, respect the ideal capacitor impedance. Ideal capacitor impedance decreases with the increase of the frequency, while the real one decreases until a certain frequency and then start to increase due to the parasitic element composition, R, L and C (Fig. 2.4).

Normally, we use capacitors on the supply voltage to have a low impedance component which is able to shunt AC noise current to ground for all the frequency range of this could be to have two different capacitors, connected in parallel, but the total capacitance is not only the sum of the two capacitor values because we have the parasitic elements.

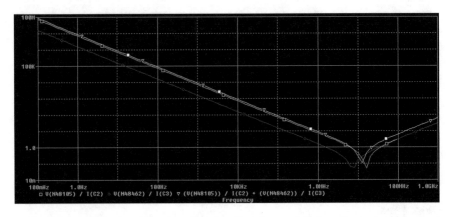

Fig. 2.5 Yellow line is the algebraic sum of the red and green. This is to show what we can wait having two parallel capacitors

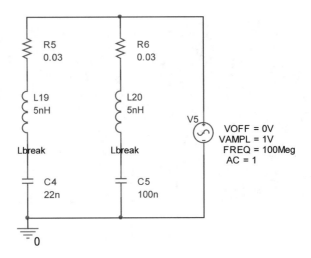

Fig. 2.6 Circuit composed by the two capacitors and their parasitic elements

Figure 2.5 shows the simply algebraic sum of the two curves, but the results of the parallel capacitor network is not so granted as shown in Fig. 2.6. The composition of the two graphs for the two capacitors acts to decrease the impedance but introduces a peak. This peak is mainly due to the parasitic resistance as shown in Fig. 2.7, and it is called anti-resonance.

Figure 2.7 shows the anti-resonance peak around 12 MHz. This peak is the result of the global circuit that is different from the sum of the single parts, and this is the normal life. This peak could be not admitted because at that frequency, the impedance could have a big value, not so low to permit a reaction versus the current change. I can personally remember a non-perfect capacitors selection on a board with the

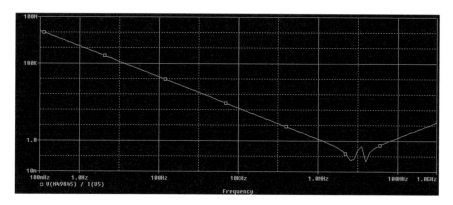

Fig. 2.7 Global impedance versus frequency of the two parallel capacitors

results to have the anti-resonance peak near to the supply voltage resonance with a dramatic result for the device. We spent two weeks to discover why, two days to change the design and two hours with the boss to be on the carpet.

It will be normally to put more than one capacitor on the supply voltage to have low impedance, but due to the parasitic elements, this is not, in general, a good suggestion. It depends from the frequency you want to manage. If your system works at 100 MHz, you will need, at least, to be able to admit a switching current on the supply node at list at 100 MHz, so the capacitors network must be able to react at 100 MHz, and the figure of merit is the impedance at that frequency; the lower the better it works.

Very often the designer adds capacitors in parallel with different values to be sure to have a good frequency response in a quite large bandwidth; a typical example you can find in a lot of circuit is made by 10 uF, 100 nF and sometime with a 1 nF.

We would like to start from this simple circuit consideration, because in the rest of the book, we will analyze other circuits, where normally, we do not always consider parasitic elements, this could be done for sake a simplicity and also because we have to write a book, not an encyclopedia.

Only about the right choosing of the capacitor on the supply voltage, it is possible to spend a lot of pages: Type of capacitors, value, how to joint to the other components, where to put in a device or in a board, etc., many and many things to be taken in account to design a robust device, but for now, we have only in mind that you need to think a lot if you want to reproduce, with a simulation, the reality.

Study what happened changing the value of the parasitic resistance, and then, do the same analysis changing the value of the parasitic inductance.

Study also what happened using capacitors with different ESR.

2.2 Supply Voltage Using a Three Terminal Adjustable Regulator

We will describe a power supply voltage using an integrated circuit, not simply transistors. This is to show that, if you have the model you can simulate also a system, the power supply voltage, having another system inside, it is only a matter of the ability to modeling different parts.

Starting by describing Fig. 2.8, the core of the design is the integrated LM317, but first, let's speak about the input transformer. The transformer we select is the ideal one. The output value is defined to assign the value of inductance of the primary coil and the secondary coil.

You have to remember what is important, for a transformer: the turn ratio.

The inductance is proportional to the square of the number of turns, and the voltage ratio from primary to secondary turn is equal to the turn ratio.

$$L \propto n^2$$

$$\frac{V_{input}}{V_{output}} = \frac{n_{input}}{n_{output}}, \frac{L_{input}}{L_{output}} = \frac{n^2_{input}}{n^2_{output}}$$

$$\frac{V_{input}}{V_{output}} = \sqrt{\frac{L_{input}}{L_{output}}}$$

With this simply rule, you can put, in the transformer model, in L1_VALUE and L2_VALUE, the value of the inductance, as shown in Fig. 2.9.

We are using a 230 Veff and 50 Hz mains supply, and we want to have 30 V output voltage.

After the transformer, we have the diodes bridge, providing a DC current starting from an AC source

Figure 2.10 shows the LM317 functional block diagram, coming from the datasheet. There is a Darlington configuration between the input and output pins.

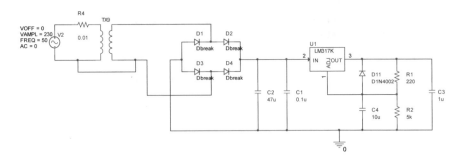

Fig. 2.8 General purpose supply voltage

		Implementation	Implementation Path	Implementation Type	L1_VALUE	L2_VALUE
1	⊞ SCHEMATIC1 : PAGE1			<none>	21H	1H

Fig. 2.9 How to obtain a transformer assigning the value of the primary and secondary coil

Fig. 2.10 LM317 functional block diagram from datasheet

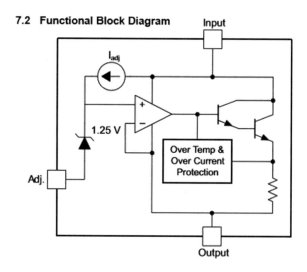

7.2 Functional Block Diagram

This is the path between the supply network and the load. This path, the Darlington, is controlled by a signal driven by a comparator that has, in input, a reference signal with the possibility to adjust it and a feedback from the output. A simple feedback network allows controlling the output voltage keeping it fixed versus the output current (load variation).

The most important technical characteristics to take in account, to design the value of the external components are summarized in the table here below.

Output voltage	Min = 1.25 V Max = 37 V
Input-to-output differential voltage dropout	Min = 3 V, Max = 40 V
Maximum output current	1.5 A
Maximum dissipated power	15 W
Output ripple rejection	64 dB

Table of the LM317 main characteristics.

The output voltage value has a minimum and maximum value. The input-to-output differential voltage dropout is a limitation to keep the device on. If you will apply, on the input of the LM317, a voltage equal to 30 V, for example, the output can has a value from 1.25 V, the minimum, to 27 V, (30 V, −3 V).

If the input voltage is greater than 40 V, like 50 V, the minimum output voltage will be 10 V (50 V, −40 V), and the maximum 47 V. The difference between the

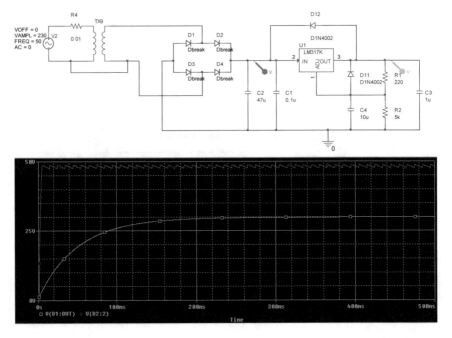

Fig. 2.11 Schematic and simulation of input and output voltages. In this case, there is no load

input and the output voltage must always be between 3 V and 40 V. The maximum current parameter defines the maximum current required by the load, and the power dissipation helps to specify the necessary heat sink.

Finally, the ripple is the residual part of the alternate power supply, −80 dB means less than 10,000 times. A lot of other characteristics are present on the datasheet.

Referring now to Fig. 2.11, where we add a diode and also the probes: C1 (0.1 μF) capacitor is connected near to the diode bridge output, while C2 is dedicated to the LM317. C2 is recommended if the regulator is not put closely to the power supply filter capacitors, to guarantee a good input signal independent from the customer layout circuit.

R_1 and R_2 set the output voltage. The producer gives a formula:

$$V_{out} = V_{ref}\left(1 + \frac{R_2}{R_1}\right)$$

With $V_{ref} = 1.25$.

If we want to have an output equal to 30 V, the input must be greater than 33 V but lower than 60 V. With R_1 equal to 220 Ω, R_2 is 5060 Ω. We can use a resistor of 5000 Ω and add, it will be better, a potentiometer to regulate, on field, the voltage.

C4 is suggested to reduce the residual ripple on the pin. The adjust pin must be stable as much as possible, otherwise you can find ripple on the output voltage.

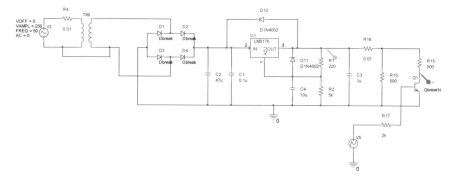

Fig. 2.12 Schematic to simulate the load change during time

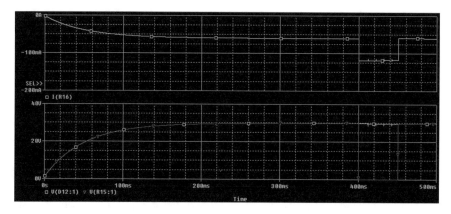

Fig. 2.13 Simulation of the schematic with load

The two diodes D11 and D12 are useful to protect the device. D11 will enter in function to discharge C4 if the output pin is shorted to ground inadvertently, same for the diode D12 to discharge C3.

Here below the simulations show how the circuit works.

Please, measure the current values in C1 and C4.

Let us see what happen with the load; simulate also what happen if the load current request changes with the time. Figure 2.12 shows the schematic.

The scheme is exactly the same but we add a probe resistor, R16, 0.01 Ω, only to verify the load current. We add the load, R10, 500 Ω and another load in parallel, R15, 500 Ω again with the Q1, an NPN simulate a switch. At certain time, Q1 will turn on and the load will become the parallel of the two resistors, and the load will became 250 Ω, with a doubling of current request.

Figure 2.13 is the simulation. The graphic below represent the output voltage and the Q1 collector, to identify when Q1 is turn on. Graphic above is the current flowing in R16. The current changes, but the output voltage remains substantially the same.

Current versus is negative, the convention! The current increases, and this explains the ripple, in the red circles.

Explore the maximum current that this circuit could deliver.

Try to change the voltage output value by changing the resistor divider value.

Analyze the stability of the circuit versus different frequency input voltage.

Is it stable the circuit versus temperature variation?

Behavior with mixed load ($R \rightarrow RC$, $R \rightarrow RL$ and $R \rightarrow RLC$).

2.3 Supply Voltage, Using Discrete Transistor, with a Negative Output Resistance

Now, we want to study a circuit done with discrete transistor, with a feedback to have like a 'negative' resistance versus the output changes. If the output decreases, because the load needs more current, the input changes its parameters driving the transistor in base mode in a way to increase the current.

As usual, we start with the schematic as shown in Fig. 2.14.

Start to describe the schematic, the input generator, V3, is the output of a transformer, 24 Veff and 50 Hz (I am in Europe, in Italy☺).

Then we will see the diode bridge, in this, case real and not ideal and finally the complete circuit. Transistor Q1, a PNP, biases the Q2, and the NPN transistor is used to modulate the input voltage to havethe desired output voltage value. Q3, NPN again, is used to implement the feedback to maintain the output in spite of the load current request change. The output voltage is across the capacitor C1.

Figure 2.13 shows that we isolate and drawn the components involved in the feedback network to establish the output value.

Across the network identified in Fig. 2.15, the Kirchhoff law to the net must be satisfied.

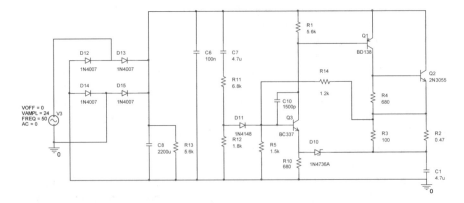

Fig. 2.14 Schematic of the supply voltage with discrete transistors

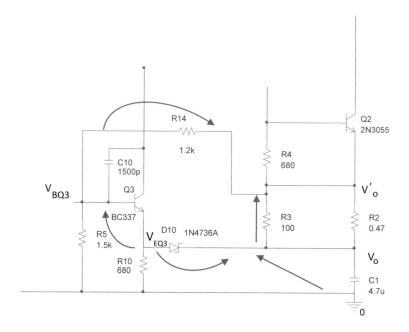

Fig. 2.15 Feedback network to determine the output voltage value

We can write, using 'engineer simplification,' with 6.8 V for the Zener threshold:

$$(V_o - V_o') - (V_o' - V_{BQ3}) - 0.6 + 6.8 = 0$$

where V_{BQ3}, thanks to the resistor divider R14, R5, is roughly equal to 0.55 V_o, and due to the fact that R3 is only 100 Ω and V_o and V_o' are almost shorted, we can rewrite the last equation as:

$$-V_o' + V_{BQ3} - 0.6 + 6.8 = 0$$

$$V_{BQ3} \cong 0.55 V_0$$

And, we can find

$$V_0 \cong 13.6$$

Strange thing could be the parallel R2 and R3. In the real circuit, R3 is a potentiometer. I used its bigger value.

Schematic and simulation, in time domain, is shown in Fig. 2.16, the circuit with the probes and then the waveforms. The feedback works in this way: If the output voltage decreases, due to the greater current request to the load, the Q3 base voltage

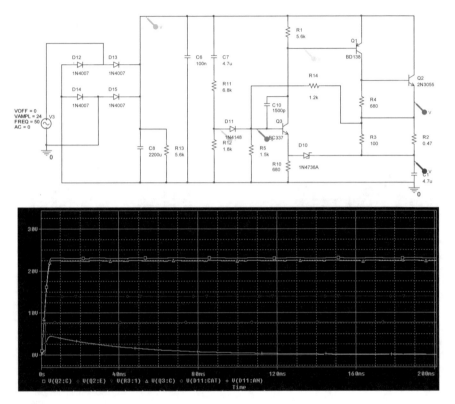

Fig. 2.16 Schematic with probes and relative waveforms

decreases and the Q1 decreases again, but Q1 is a PNP; so, the Q1 collector voltage, common to the Q2 base, increases, and this increases the output current, as requested by the load.

An interesting item to be analyzed is the role of the C7, R11 and R12 (plus the diode D11). As you can see in the last simulation, waveform light blue is a one pulse, only to the start, to turn on the Q3 transistor, and when the circuit starts, after this moment, the circuit is self-sustaining.

A second remarkable component is the capacitor C10. It is a short between collector–base junction higher frequency with respect to the 50 Hz of the supply network.

Without C10, the feedback circuit could be not so fast to recover the change, introducing an oscillation on the output voltage.

A suggestion is to proceed like we did in the previous paragraph, add a load, a resistor and a switch and simulate what happens.

Study the feedback network.

Reduce and remove the C10 and then observe what happens.

Study the output voltage versus output current to evaluate the load regulation.

2.4 Switching-Mode Power Supply Theory and Practice

Now, a different concept: If you have a direct current generator (DC current) and you need to transform the value of the voltage up or down, with respect to the voltage you have, a switching supply is useful.

This real different way to work shows the incredible variety provided by the different combinations of electronic components.

In the previous paragraphs, we work with supplies voltage based on a transistor that is able to sustain a voltage drop to limit the output voltage. We have a greater input voltage, the output of the transformer, and we have a low voltage output where the voltage difference is sustained by a transistor junction.

This transistor dissipates, by Joule effect, this excess power with efficiency down grade.

The transistor works like a variable resistance in series with the load.

The output voltage, therefore, will be less than the input voltage. With this type of the supply generator, the efficiency could be also lower than 60% and normally needs a heat sinker. A switching could reach efficiency greater than 90% and not need to dissipate because, thanks to the high efficiency, there is no heat to dissipate, or less with respect to the other solution.

The cons are a residual ripple, difficult to eliminate and to be taken in account if we do not want to have 'noise' on the output, like in the hi-fi supply.

Before studying a switching power supply using an integrated circuit, we will start with a more theoretical analysis for a step-down and a step-up supply converter.

Figure 2.17 shows a typical step-down configuration. The value of the components was assigned using a simple algebra that we cannot reproduce here, we suppose the reader already know or, in any case, you can obtain them from simple analysis. A really helpful Application Note document is indicated in [3].

The circuit is supplied by a 20 V DC generator.

To understand how the circuit works, we have to separate the two situations—when the switch represented by the transistor is ON and when the transistor is OFF.

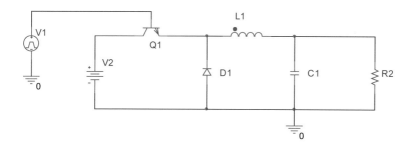

Fig. 2.17 Conceptual schematic for the step-down switching configuration

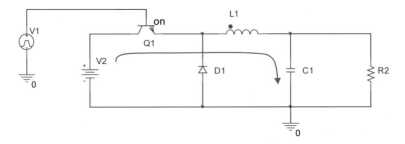

Fig. 2.18 Step-down Q1 on phase

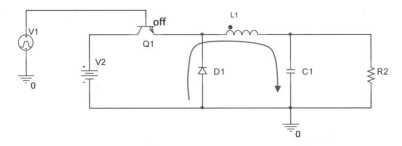

Fig. 2.19 Step-down Q1 off phase

When the transistor is ON, the current flows in the inductor to charge the capacitor and diode has a reverse bias. The current flows also in the load R2, of course, but imagine that, R2 is big and the current sink is negligible as shown in Fig. 2.18.

When the switch is OFF, the transistor is OFF, the current in the inductor will flow in the same verse, like in the previous phase, thanks to the diode, the inductor 'will discharge the magnetic field' stored in the previous phase charging again the capacitor, Fig. 2.19.

The output voltage depends, mainly, from the ratio between the Ton and Toff switch time. We can say

$$V_{\text{out}} = V_{\text{in}} \frac{T_{\text{on}}}{T_{\text{on}} + T_{\text{off}}}$$

Example, with a V_{in} equal to 20 V and a duty cycle of 25%, we will have a V_{out} equal to 5 V.

The 'secret' of the switching is that the power dissipation is very low; during the ON phase, the voltage drop on the transistor is equal to its saturation voltage, generally very low, but with high current, when it is in the OFF phase, the voltage is high, but the current is zero. In both situations, the power, the product voltage by current, is very low.

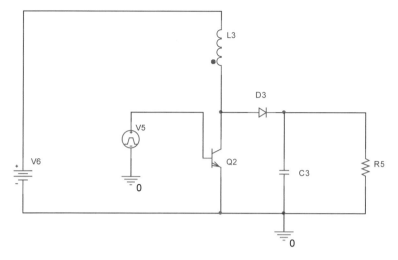

Fig. 2.20 Conceptual schematic for the step-up switching configuration

Second type of configuration that we consider is the step-up configuration, and how to obtain high voltage output with respect to the voltage input is represented in Fig. 2.20.

Also, in these cases, the value to the components could be found by the reader. Many times I hate the authors when I found these types of sentences, but please remember, we suppose that the level of knowledge of the reader is high, otherwise, like you can discover through the rest of book, the number of the circuits topology treated is too large to be too much detailed.

The purpose is to show how to test circuits with a 'virtual laboratory' analysis.

So, when the transistor is ON, the current flows in the inductor to ground charging the inductor as shown in Fig. 2.21, and when the transistor switches OFF, the inductor maintains current direction to charge the capacitor as shown in Fig. 2.22. The next ON phase will charge again the inductor, but the capacitor will maintain its voltage because the diode is reverse biased.

In this case, the output voltage could be calculated, in a simple manner, as:

$$V_{out} = V_{in} \left[\frac{1}{1 - \left(\frac{T_{on}}{T_{on} + T_{off}} \right)} \right]$$

If the input supply is 9 V again, with a duty cycle of 68%, like in the step-down example, we will have a V_{out} equal to 28 V. Of course, we have to take in account the current, not only the voltage, so we have to add the load.

Study the current available for the load, in the step-down and the step-up circuit.

Let us now start to simulate a more realistic circuit using an integrated circuit device, designed to generate all the signals we need.

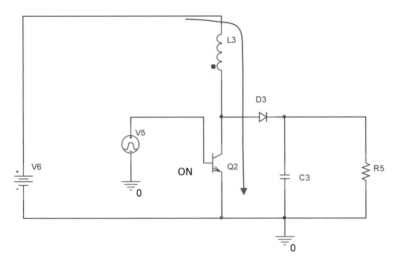

Fig. 2.21 Step-up Q1 ON phase

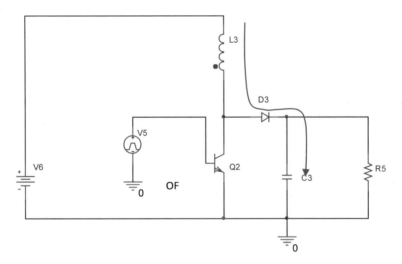

Fig. 2.22 Step-up Q1 OFF phase

Figure 2.23 shows the block diagram for the MC34046A device. Block diagram components are, of course, very similar to the linear supply voltage generator that we see in Fig. 2.10, but the way to use the components is different. Darlington couple is the switch and not, like in the linear version, the element that was used to drop the voltage. The comparator and the reference generator are the same, and in qualitative way, we have the oscillator like a new component, to produce the switch transistors ON and OFF. In the relative component datasheet, it is possible to find what is reported in Fig. 2.24, where the producer explain how to calculate the value of the

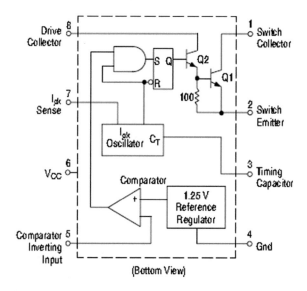

Fig. 2.23 MC34046A schematic diagram

Calculation	Step–Up	Step–Down	Voltage–Inverting
t_{on}/t_{off}	$\dfrac{V_{out} + V_F - V_{in(min)}}{V_{in(min)} - V_{sat}}$	$\dfrac{V_{out} + V_F}{V_{in(min)} - V_{sat} - V_{out}}$	$\dfrac{|V_{out}| + V_F}{V_{in} - V_{sat}}$
$(t_{on} + t_{off})$	$\dfrac{1}{f}$	$\dfrac{1}{f}$	$\dfrac{1}{f}$
t_{off}	$\dfrac{t_{on} + t_{off}}{\dfrac{t_{on}}{t_{off}} + 1}$	$\dfrac{t_{on} + t_{off}}{\dfrac{t_{on}}{t_{off}} + 1}$	$\dfrac{t_{on} + t_{off}}{\dfrac{t_{on}}{t_{off}} + 1}$
t_{on}	$(t_{on} + t_{off}) - t_{off}$	$(t_{on} + t_{off}) - t_{off}$	$(t_{on} + t_{off}) - t_{off}$
C_T	$4.0 \times 10^{-5}\, t_{on}$	$4.0 \times 10^{-5}\, t_{on}$	$4.0 \times 10^{-5}\, t_{on}$
$I_{pk(switch)}$	$2I_{out(max)}\left(\dfrac{t_{on}}{t_{off}} + 1\right)$	$2I_{out(max)}$	$2I_{out(max)}\left(\dfrac{t_{on}}{t_{off}} + 1\right)$
R_{sc}	$0.3/I_{pk(switch)}$	$0.3/I_{pk(switch)}$	$0.3/I_{pk(switch)}$
$L_{(min)}$	$\left(\dfrac{V_{in(min)} - V_{sat}}{I_{pk(switch)}}\right)t_{on(max)}$	$\left(\dfrac{V_{in(min)} - V_{sat} - V_{out}}{I_{pk(switch)}}\right)t_{on(max)}$	$\left(\dfrac{V_{in(min)} - V_{sat}}{I_{pk(switch)}}\right)t_{on(max)}$
C_O	$9\,\dfrac{I_{out}t_{on}}{V_{ripple(pp)}}$	$\dfrac{I_{pk(switch)}(t_{on} + t_{off})}{8V_{ripple(pp)}}$	$9\,\dfrac{I_{out}t_{on}}{V_{ripple(pp)}}$

V_{sat} = Saturation voltage of the output switch.
V_F = Forward voltage drop of the output rectifier.

The following power supply characteristics must be chosen:

V_{in} – Nominal input voltage.
V_{out} – Desired output voltage, $|V_{out}| = 1.25\left(1 + \dfrac{R2}{R1}\right)$
I_{out} – Desired output current.
f_{min} – Minimum desired output switching frequency at the selected values of V_{in} and I_O.
$V_{ripple(pp)}$ – Desired peak–peak output ripple voltage. In practice, the calculated capacitor value will need to be increased due to its
equivalent series resistance and board layout. The ripple voltage should be kept to a low value since it will directly affect the
line and load regulation.

Fig. 2.24 How to define component values

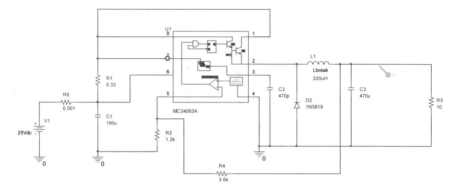

Fig. 2.25 Application step-down configuration

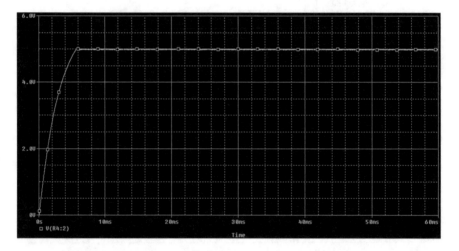

Fig. 2.26 Step-down configuration voltage output

component and the electrical parameters you need [4]. This time, we will consider
the application scheme in the datasheet, Fig. 2.25, and we redesign it with our tool.

R6 is with a very low value resistor added to measure the current and then to
calculate the power efficiency; the load is R5, 10 Ω.

As we can see in the simulation, Fig. 2.26, the output voltage is 5 V, like we want,
coming from 25 V input generator, V1. This means 500 mA is on the load.

Calculated the efficiency for this circuit.

Figure 2.27 shows, for the same device, the step-up configuration and in Fig. 2.28
the simulation with the output voltage: 28 V, starting with a voltage input equal to
12 V.

Also, for this case, calculated the efficiency of the circuit.

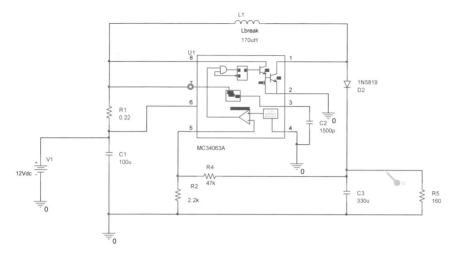

Fig. 2.27 Application step-up configuration

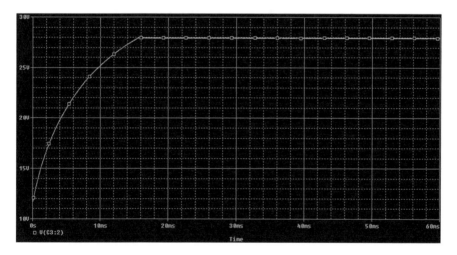

Fig. 2.28 Step-up configuration voltage output

2.5 Transformerless Power Supply

A very interesting way to design a supply voltage is used also in a domestic light dimmer, without the transformer. A simple configuration is shown in Fig. 2.29. The idea is to connect 'almost' directly the supply network at 230 V, 50 Hz to the diode bridge. We use a capacitor, between the 230 V and the diode bridge, to decrease the voltage and then a Zener diode to stabilize the output voltage. With this solution, we can avoid the transformer; this could be a dangerous solution because in case on

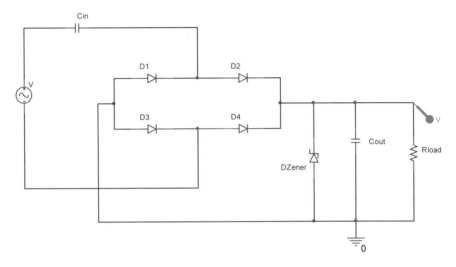

Fig. 2.29 Transformerless schematic solution

short circuit or failure occurring to the circuit, the 230 V may, inadvertently, come into contact with the user. For this reason, great care must be taken to isolate the circuit from all the possible user interactions.

Let me show few formulas to explain how to analyze the circuit. First assumption, to do the next calculation is that the value of the output voltage is negligible (for an engineer means less then) with respect to the input voltage.

The current flowing in the input capacitor is:

$$I_{peak} = \omega C_{in} V_{peak}$$

While the current flowing in the load is:

$$I_{out} = \frac{V_{out}}{R_{load}}$$

With the hypothesis that the average current flowing in the input capacitor is equal to the load current, we can write

$$\frac{V_{out}}{R_{load}} = \frac{2}{\pi} \omega C_{in} V_{peak} = \frac{2\sqrt{2}}{\pi} \omega C_{in} V_{in}$$

From this equation, it is easy to verify that

$$V_{out} = 4\sqrt{2} f C_{in} R_{load} V_{in}$$

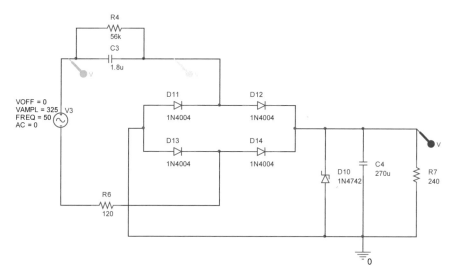

Fig. 2.30 Complete transformerless schematic to have 12 V from domestic European network

And finally,

$$\frac{V_{out}}{V_{in}} = 4\sqrt{2}\,f\,C_{in}\,R_{load}$$

And also,

$$I_{out} = 4\sqrt{2}\,f\,V_{in}$$

Rule of thumb, for a 230 V, 50 Hz supply network, we can have 65 mA for each microFarad of input capacitor.

A complete circuit is designed in Fig. 2.30.

The difference from Fig. 2.29 is the two resistors, R4 and R6. R6 is needed to limit the Zener diode current, while R4, with C3, is a filter to limit EMI disturb coming back on the line.

Simulation is depicted in Fig. 2.31, and the component values were designed to have 12 V on the output, blue probe [5].

The reader could try to design the value of the resistors, the capacitors in terms of not only value but also defining the power characteristic.

Verify what happen if the input capacitor C13 breaks.

Verify the current value available on the load.

Check the ripple obtained.

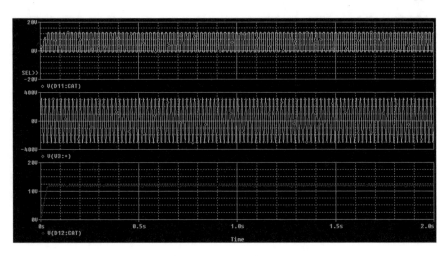

Fig. 2.31 Input–output voltage simulation for the transformerless

2.6 Something About Power Design Network (PDN)

We would like to close this chapter giving an overview of the power network design. When your design is complex, works at high frequency and has a big power consumption, the design of the power supply network must be done using more accurate techniques. We open the chapter talking about capacitors normally connected to the supply as decoupling. Now, we will show some criteria for a more realistic power design.

First point is to understand the element composing a power network. A device, an integrated circuit like a microcontroller, is put in a package, and it is connected, soldered on a board and to be connected to a power supply; we have PCB board lines or wires.

Figure 2.32 shows the connections done, of course, not only by components but also by parasitic. Starting from the left side, we have power supply, the ideal generator and its internal resistance, V_{supply} and R_{supply}, the RL cable parasitic then the printed circuit board (PCB) RL parasitic and then the contribution of the 'bulk' capacitors and the decoupling capacitors, the power plane contribution, the package connection; imagine a wire bonding device, and finally, the device is represented by a variable resistance, what we call 'the load.'

A note about 'bulk' capacitors is a way to distinguish capacitors that dedicate to the supply voltage device, normally having a big values, used like a tank and the decoupling capacitors connected as near as possible to the different integrated circuit to limit the parasitic connection. At the end is the same concept but managed, many time, by another designers.

Figure 2.33 shows a possible network with a common value for parasitic and components, in reality will be, of course, more complex.

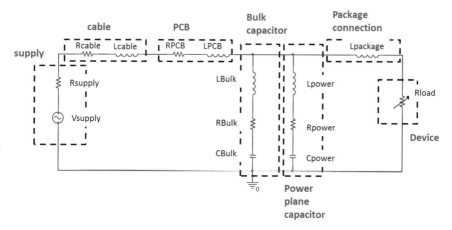

Fig. 2.32 Conceptual typical power network elements

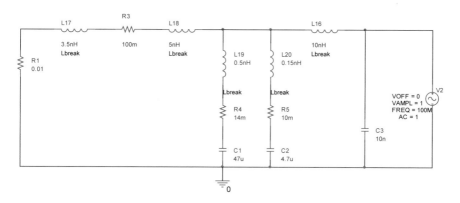

Fig. 2.33 A realistic network for a device assembled on a PCB board

In this situation, we remove the device, the load, and substitute it with a generator because we want to calculate how the impedance, of the supply network, reacts to different frequency, we suppose that it will be generated by the different current requested from the load.

Figure 2.34 shows the simulation, the impedance views from the load side, versus the frequency.

In our design, if we stay under 1 MHz, the impedance is flat and reasonable, less than 100 mΩ. If the load start to work at high frequency, the impedance raises and reaches 125 Ω at 16 MHz.

Of course, this could not be acceptable, and the designer must take action to change the supply power network.

The designer always pursuit the flatness of the impedance characteristic and not only, he would like also to have a precise value by design. This value could be defined

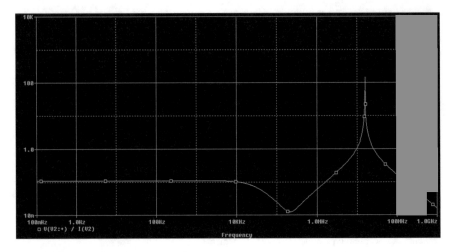

Fig. 2.34 Network impedance versus frequency

asking to have a minimum ripple of the supply voltage node with respect to the load current variation.

We can define:

$$Z_{\text{target}} = \frac{V_{\text{supply}} \cdot \text{ripple}\%}{I}$$

In our case, if we have, for example, V_{supply} equal to 3.3 V, a current required by the load equal to 2 A and a ripple admitted equal to 7%, the result is:

$$Z_{\text{target}} = \frac{3.3 \cdot 7\%}{2} = 115.5 \text{ m}\Omega$$

This is the target value, of course, this is the best value we can have, and we also would like that this remain flat over the frequency range.

After this analysis, we have to take in account another important thing. Until now, we analyzed the behavior of the power network, but we have to consider that the load, we have simulated this with a variable resistor, varies with a different ways, not always like a pure sinusoidal wave but, of course, with a signal composed by a lot of different sinusoidal waves. In other term, we need to understand how to 'really' simulate the load current variation to understand if the power network we plan to use is good enough to sustain the load current requests.

We cannot go more deeply in this problem, but it is important to remember that the digital circuit is normally more difficult to consider because, could be seems strange, like many other things, but, a digital circuit, due to the sharp edge in the signal commutations, has a frequency content more complex than an analog circuit. Figure 2.35 shows a simple schematic. On the left, a load, a resistor, connected to a sinusoidal generator, represents an analog circuit. On the right side, e same circuit

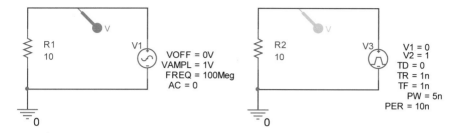

Fig. 2.35 Left represents an analog circuit, right a digital circuit

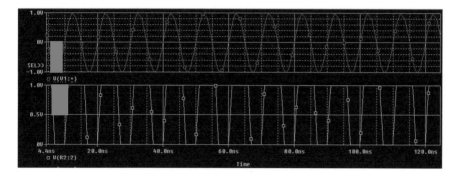

Fig. 2.36 Analog and digital signal in the time-domain example

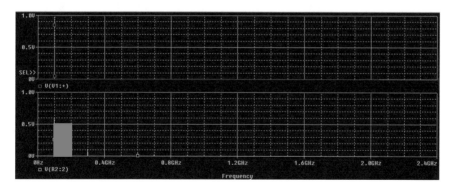

Fig. 2.37 FFT for the previous analog and digital signals

but with a digital generator, to have a digital signal like a clock, fast edge to rise and fall, time to have a stable signal high and low level.

Figure 2.36 shows the signals 'analog' and 'digital' in the time domain.

Figure 2.37 is the fast Fourier transform analysis done on the two previous signals. It is simple to see that, the analog, in this case, we use a pure sinusoidal wave, which has only a frequency, and in the digital situation, we have a certain number of frequency contributions; so, we have to imagine, for our load of Fig. 2.32, a very complicated frequency contents and we have to satisfy, with our power design network, to have a stable and well-controlled system, working at all the frequency defined.

Today, due to the big devices produced, with billions of transistors, working at more at GHz, with a power consumption of more than 5 A, the power design network is a must. Many dedicated tools exist to perform the flow that we showed to extract parasitic values and perform simulation at high frequency [6, 7].

Chapter 3
Linear Amplifiers

3.1 Introduction

In this chapter, we will make some experiments with linear voltage amplifiers using BJT transistors even if most of the consideration we will make can be applied also to JFETs or MOSFETs.

We will consider first a simple two-stage amplifier looking at the most important features of the schematic. We will see the applied concept of *feedback* and of *compensation* demonstrating how the gain and bandwidth of the amplifier can be determined by few resistors and capacitors independently as much as possible from active devices characteristics.

We will later consider some special circuit building blocks commonly used in the linear circuit design: Darlington configuration and current mirrors. Then, we will analyze main blocks of a simplified circuit to understand some concepts used in the design of linear amplifiers.

Finally, an example of a power audio amplifier will be given.

3.2 Multistage Amplifiers

Certainly, during your school time, you learned the circuit of a single transistor amplifier and studied the DC bias, the voltage gain and the behavior in frequency. Today, I would consider a two transistor amplifier instead, like the one shown in Fig. 3.1.

It is a two-stage transistor amplifier that could be used as a preamplifier of the signal coming from a signal source in an audio reproducing system.

The amplifier is made of a first transistor (Q_1) operating in common emitter configuration, with a feedback emitter resistor to stabilize the voltage gain, while the proper bias to the transistor is taken through the network R_5–R_7 that at the same time provides a feedback path for stabilization of operating point of the whole circuit.

© Springer Nature Switzerland AG 2020
R. Gastaldi and G. Campardo, *Electronic Experiences in a Virtual Lab*,
https://doi.org/10.1007/978-3-030-45179-0_3

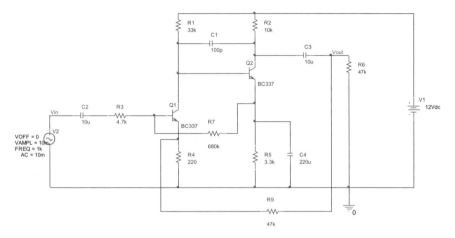

Fig. 3.1 Schematic of a two-stage linear amplifier

This feedback is active only in DC mode because of the shunt capacitor C_4, and it is necessary because the two stages are DC coupled. From base and collector of Q_2, it is connected a capacitor C_1 in parallel with collector-base capacitance of Q_2. The signal flows from first stage to the second one with a direct connection, without a decoupling capacitor. Instead, a decoupling capacitor C_2 is connected between the input signal and the amplifier circuit, and another capacitor C_3 is connected before the output load resistor R_6.

In this way, the DC operating point is not affected by an eventual DC component of the input signal source and an eventual offset cannot affect the load.

To increase the linearity of the amplifier, a negative feedback from output to input has been added using R_9. This network senses the output voltage of the amplifier, across resistor R_6 and uses this information to modify voltage at the emitter of Q_1 using resistor R_4. We will discuss this network later.

Let's now launch a simulation focusing on bias conditions when input is zero volt. Voltages in all the nodes of the circuit resulting from simulation are shown in Fig. 3.2, but before going on, I suggest you to make an exercise trying to answer the following questions:

- *Why did we choose a direct coupling between the two stages?*
- *We said before that a bias stabilization feedback is heavily recommended in this case. Do you guess why?*

We can now simulate the dynamic behavior of the circuit, using the sinusoidal generator V_2. I have chosen a frequency of 1 kHz because at this value, the shunt capacitor C_4 and decoupling capacitors C_2 and C_3 can be for sure considered a short circuits. Simulation results are shown in Fig. 3.3 where we can see that the voltage gain of this amplifier is about 200. I could calculate this gain from the analysis of the circuit using the hybrid-pi parameters model for example, but if I suppose that

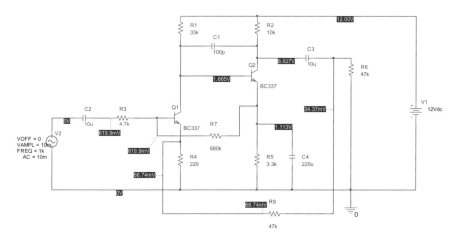

Fig. 3.2 Circuit of Fig. 3.1 in static conditions. Bias in every node of the circuit is shown

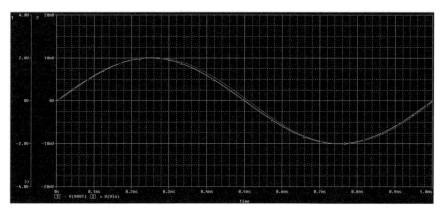

Fig. 3.3 Response of the amplifier (in red) to the application of a sinusoidal signal (in green) of amplitude 10 mVpp and 1 kHz frequency

the gain of the amplifier without feedback (A_v) is very large, remembering that from the feedback theory the voltage gain of an amplifier with feedback A_{vf} is given by:

$$A_{vf} = \frac{A_v}{(1 + \beta A_v)} \tag{3.1}$$

With β representing the feedback gain, it comes out that:

$$A_{vf} \cong \frac{1}{\beta} \tag{3.2}$$

We can notice that the amplifier is not-inverting and comparing the Y-axis scale of the input and output signals that are almost superimposed in the figure the voltage gain is 200. Input signal amplitude has been chosen small enough to maintain the system in the linear zone so that small-signal analysis method is valid.

Let's now see what happens if we calculate the voltage gain using theory of feedback amplifiers: We know that for this kind of feedback configuration feedback gain is given by:

$$\beta = \frac{R_4}{(R_4 + R_9)} \tag{3.3}$$

Using the numbers of our schematics $\frac{1}{\beta} = (47{,}000 + 220)/220 \cong 214$
Which is reasonably near the value we can see from Fig. 3.3.

To better check this result, I suggest you to change the value of R_9 and observe through simulation how the gain changes.

Now, it's time to make some work about the bandwidth. It is obvious that this is a very important figure of merit in a linear amplifier. Let's use a very useful feature of OrCad that performs a frequency sweep of the input waveform generator making a simulation at every frequency step and plots directly the output voltage as a function of frequency in a logarithmic scale. Let's put the start frequency of the sweep at 10 Hz and the stop at 30 kHz and run the simulation …

From the figure, we can see that the voltage gain of the amplifier is almost flat up to 3 kHz then starts to roll-off. Usually, the upper cut frequency of the amplifier is defined as the one at which the gain has been reduced of a factor 0.707. Looking at Fig. 3.4, we can say that the cut frequency of this amplifier is about 20 kHz.

In the figure caption, I specified the value of C_1 used in the simulation. Before talking about the role of C_1, I would like to use this circuit to make an exercise to understand better the theory of linear amplifiers. This should give us a better insight

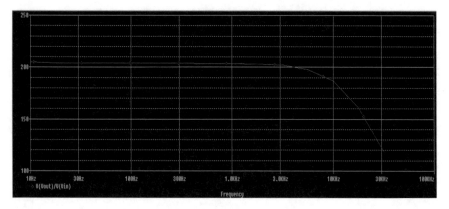

Fig. 3.4 Behavior of the gain $= V_{out}/V_{in}$ with frequency of the two-stage amplifier with $C_1 = 100$ pF

Fig. 3.5 Hybrid-π model
for a BJT

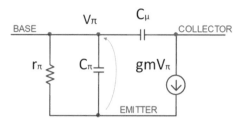

about how the circuit works and in some way we can say that it could be used even
if we don't have a simulator available. In fact, we will not use the models of active
devices embedded in the software but the small-signal model shown in Fig. 3.5;
instead, we will still use the simulator to quickly solve the resulting network.

This circuit is the so-called hybrid-π model for a BJT transistor. Here, it has been
simplified by neglecting the output resistance that should take into account the early
effect. The first thing to do is to determine the DC operating point of the transistor,
because as you know the parameters of the model can be calculated and the model
is valid for 'small' signals around the operating point.

Once again, we can use our simulator as a measuring instrument to achieve the
DC operating point of the circuit. The result is shown in Fig. 3.6. Now, it is possible
to calculate the parameters for the transistor small-signal model using the definitions
that I recall here below:

$$r_\pi = \frac{h_{fe}}{g_m} \tag{3.1}$$

$$g_m = \frac{q}{KT|I_C|} \tag{3.5}$$

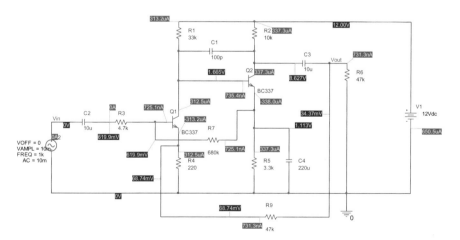

Fig. 3.6 DC operating point of the two-stage amplifier showing voltages (brown) and currents (red)
in every node

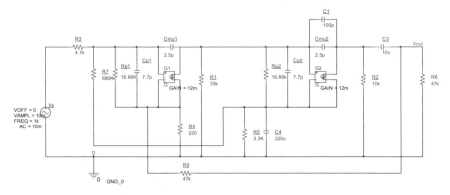

Fig. 3.7 Small-signal equivalent circuit of the amplifier

$$C_\pi = \frac{g_m}{\omega_T} - C_\mu \tag{3.6}$$

$$C_\mu = C_{ob} \tag{3.7}$$

where I_C is the DC collector current, h_{fe} is the DC current gain, ω_T is the cutoff angular frequency and C_{ob} is the output capacitance of the transistor that can be found in the transistor datasheet.

Now, it is possible to draw the complete small-signal model that is reported in Fig. 3.7.[1] Please note that, I have used the hybrid-π model in place of BJTs and I have put to zero the DC voltage generator V_1. *Make a trial and simulate this circuit and compare the result with Fig. 3.4 and you should find about the same value for gain and the roll-off curve.*

What we want to do now is to verify the relationship (3.1) using the small-signal network we have just built.

First of all then, at the frequency we are operating, let's consider as short circuits the coupling capacitors C_2, C_3 and the shunt capacitor C_4 which, just to remember, allows to exclude the DC bias stabilizing feedback of R_4–R_7 during AC operation.

As a second step, we break the feedback loop separating the gain amplifier stage A_v from the feedback network β done with R_4–R_9. The stage without feedback but taking into account the load generated by the feedback network is shown in Fig. 3.8.

You should notice that in place of R_4, R_9 we have now, respectively, R_{4a}, R_{9a} for the input loop and R_{4b}, R_{9b} for the output loop.

Using this circuit, you can launch a simulation to find the gain (A_v) of the circuit without feedback. Let's do this and look at the results in Fig. 3.9. We can notice that gain is very high at low frequency, about 2800 compared with 200 of the feedback amplifier but starts to decrease from 300 Hz and becomes almost zero at about 1 MHz.

---

[1] The components of hybrid-π model in Fig. 3.7 have a different name with respect to Fig. 3.5 where I have used the names you find in every text-book, but they can be easily recognized.

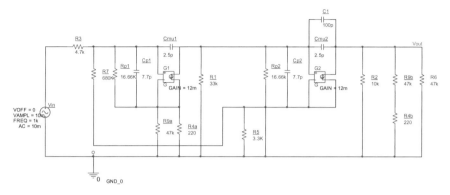

Fig. 3.8 AC small-signal equivalent circuit of the amplifier: coupling and shunt capacitors are shorted and AC feedback broken, but the loading effects of the feedback network are included

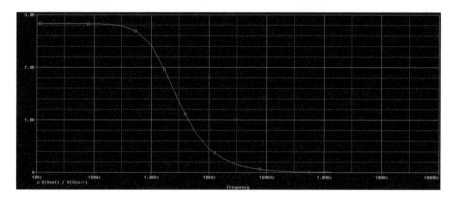

Fig. 3.9 Voltage gain of the equivalent amplifier without feedback as a function of frequency

We also know from theory that the transmission factor (β) of the feedback network is given by (3.3). Then, for low frequency in the zone where the gain is flat, we can write:

$$A_{vf} = \frac{2800}{\left(1 + \frac{2800}{214}\right)} = 198$$

which gives less than 10% difference with the approximate value found with (3.2).

Now, I would like to focus on capacitor C_1. To understand the role of C_1, it is necessary to refer to the concept of stability of the system. You know that a negative feedback amplifier can become unstable if the magnitude of the loop gain $|A\beta| > 1$ when the phase angle of $A\beta = 180$. We can use our model to verify if this situation is present at some frequency. We can plot again our loop gain in dB

$$A_v\beta(\text{dB}) = 20 \times \log_{10}|A_v\beta| \tag{3.8}$$

and $A\beta$ phase, let's see these results in Fig. 3.10. The point to look at is the value of the phase when $A_v\beta$ (dB) $= 0$ or in other words when $A_v\beta = 1$. We can see that when the phase shift is 180°, the loop gain is well below 1 and oscillations are not possible.

Let's now repeat this procedure after removing C_1 from the schematic, you will find the result shown in Fig. 3.11.

From Fig. 3.11, you can see that the bandwidth of the circuit is larger but the gain and phase margins are now reduced.

C_1 is called a compensation capacitor, and it introduces a dominant pole in the circuit, so that the upper limit of the bandwidth is easily controlled by a component independent from the device characteristics which are subject to statistical variations bringing instability to the system.

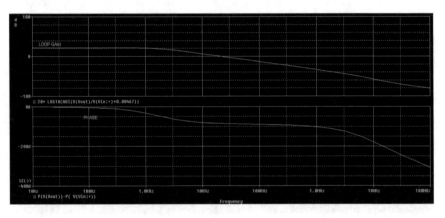

Fig. 3.10 Loop gain and phase of the circuit of Fig. 3.8 as a function of frequency with a compensation capacitor C_1

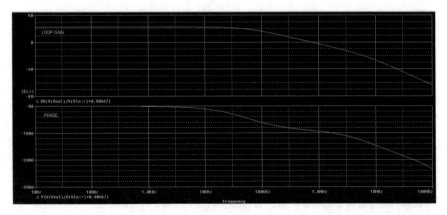

Fig. 3.11 Loop gain and phase of the circuit of Fig. 3.8 without a compensation capacitor C_1

3.3 Darlington Configuration

Before proposing to you a new circuit to analyze, I would like to spend some words on some circuit configurations that can be considered as a sort of *building block* of more complex circuits. The reason why I would like to recall these circuits here is that they are so widely used in the common practice of linear circuits design that often their operation principle is considered as known matter.

The first circuit we will see is called *Darlington configuration*, and it is shown in Fig. 3.12.

The circuit depicted in this figure can be considered a three terminal device which has a current gain

$$G = \frac{I_{e2}}{I_{b1}} \cong \beta_1 \beta_2 \tag{3.9}$$

In Fig. 3.12, it has been added the resistor R which allows to bias the first transistor with a current higher than the one flowing into the base of the second one so that a higher β is obtained.

The Darlington configuration has an high input impedance, in the order of

$$R_1 > \frac{\beta_1 \beta_2}{g_{m2}} \tag{3.10}$$

Then, it combines an high input impedance, like an FET device, and an high transconductance typical of a bipolar transistor (BJT), and it is used as a decoupling stage or voltage gain stage.

Fig. 3.12 Schematic of a
Darlington configuration

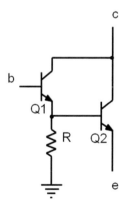

3.4 Current Mirror

Often in the design practice, it is necessary to generate a reference current and to copy it or a multiple of it in different points of a circuit. In particular, this is necessary in linear integrated circuits design when it is necessary to properly bias a node of the circuit without using integrated resistors. If you think that using an integrated resistor is a simpler try to design an integrated operational amplifier, you will see how much chip area you consume for them and how much you have to take care of process variations and layout geometry impacting resistance value, without mentioning temperature variations!

The most diffused and simple circuit to generate a reference current and to reproduce, or 'mirror' it is the current mirror shown in Fig. 3.13.

A reference current is generated forcing a fixed voltage across the resistance R, due to the fact that base and collector of Q_1 are shorted, so that $V_{\mathrm{CE1}} = 0.7$ V and we can write

$$I_{\mathrm{ref}} = \frac{(V_{\mathrm{cc}} - 0.7)}{R} \qquad (3.11)$$

If Q_1 and Q_2 have the same current gain (β) and are at the same temperature, the collector current of $Q_2 = I_U$

$$I_U = \frac{I_{\mathrm{ref}}}{\left(1 + \frac{2}{\beta}\right)} \qquad (3.12)$$

The current mirror configuration is widely diffused in the design of monolithic integrated circuits because in that case it is possible to make two active devices with a predefined β. We could demonstrate that in the circuit of Fig. 3.13a if $\beta_2 = n\beta_1$, then $I_U = nI_{\mathrm{ref}}$, so having the control of β, it is possible to build different reference

Fig. 3.13 Current mirror: **a** using bipolar transistors, **b** using JFET/MOS

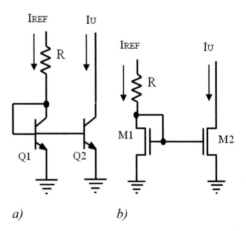

currents related to a single reference source. This is also a diffused practice in design of linear IC's.

Other reference current generators are shown in the following: The circuit shown in Fig. 3.14 is called 'Widlar mirror.'

In this circuit, you can have the following relationship

$$I_U \cong I_{\text{ref}} \frac{R_2}{R_1} \tag{3.13}$$

Another common configuration is the Wilson current mirror shown in Fig. 3.15.

A small exercise for you is to evaluate the ratio $\frac{I_1}{I_2}$ and compare it with the result (3.12). *What you can conclude?*

Fig. 3.14 Widlar current mirror

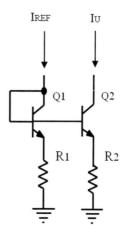

Fig. 3.15 Wilson mirror

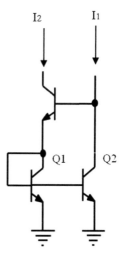

Current mirrors are often used as load elements for differential pairs amplifier stage as you can encounter, for example, in integrated operational amplifier. But I understand I am going around too much with theory and it's time to amuse ourselves with another amplifier.

The circuit I would like to propose now is a sort of example to familiarize with some concepts that are present in most of the amplifier design practice.

3.5   An Example of Linear Amplifier

The circuit of this amplifier is shown in Fig. 3.15. This schematic doesn't represent an optimized real design, so active devices used are generic NPN and PNP and diode devices.

The first thing you can notice in the schematic is the high number of active devices. This is more common in IC's design where often is easier to use, for example, active devices in place of resistances, for example, as an active load or in current mirrors to create bias currents to be used in the different blocks of the circuit. I have distinguished three main blocks that are recurrent in the design of a generic linear amplifier.

In the red circle, it is contained that the input stage which provides amplification of the input signal. In this case, it is made of the emitter coupled pair Q_2–Q_4. Together with the load Q_1–Q_3, they form a differential amplifier providing voltage gain. I assume you know everything is necessary of a differential coupled pair. You know that the two emitters of the pair must be biased with as much as possible a current generator, and this is exactly the role of Q_{15}. In this case, instead of two resistors, still a current mirror is used as a load of the differential amplifier to increase the gain of the stage thanks to its high output resistance. In fact, you know that the differential voltage gain of this stage G_d is given by

$$G_d = \pm g_m \frac{R_L}{2} \tag{3.14}$$

Q_1, Q_3 can also ensure that the bias current in both the legs of the coupled pair is equal, so the stage is completely balanced. This is very important to guarantee the good operation of the circuit.

We know that the differential stage needs a constant current bias at the common emitters, and this is ensured by Q_{15} connected as a leg of a current mirror. The output signal is taken in single-ended mode from the collector of Q_4.

In green, you can see what I called output stage that actually includes also a driver stage made of Q_7, Q_8 that are connected in Darlington configuration to provide a further gain stage. The load of this driver is a little complex including two diodes D_1, D_2 and Q_{17} working as a current generator. The output complementary pair Q_{12}–Q_{13} works in such a way that each one BJT is in conduction only during the positive and negative part of the input signal, respectively. This method allows a considerable saving of power during operation, and of course, it makes sense when high-power

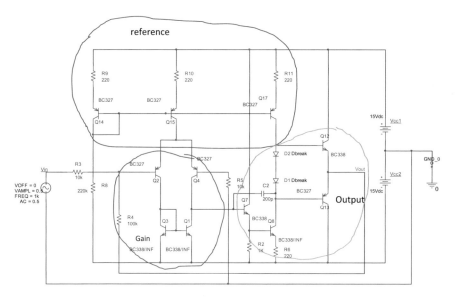

Fig. 3.16 Schematic of an example amplifier design. Three main blocks are identified

amplifiers are considered, otherwise it is not used (except in this example) due to the more critical biasing circuitry. We will come back to this point.

The last block you see blue-circled in Fig. 3.16 is named as reference because inside it are generated all the bias currents used here and there in the circuit. Basically, it is a current mirror where the reference leg is made by R_9–Q_{14}–R_8. This current is then 'mirrored' in the other legs, but the presence of R_9–R_{10}–R_{11} allows to change the ratio between the reference current and their mirrors according to the design needs.

Now, let's enter the schematic and simulate our circuit using the usual 1 kHz sinusoidal input signal. We should find the result as shown in Fig. 3.17.

Please note that the input source is coupled in DC mode, and then, input signal offset was put to zero to avoid to deal with a DC component. Also, a symmetrical power supply has been chosen to maintain the output at zero in DC condition. Actually, the feedback acts to keep the two inputs of the differential stage almost at the same voltage except for the error ε, which in turn results in zero voltage at the output node.

We can see from Fig. 3.17 that the gain of the amplifier is 10, an output of 5 V for an input of 500 mV, and it is inverting. If we look at the schematic, we see that we have put a resistance R_4 from the output node to the input node of the amplifier while R_3 is in series with the input source. This is the same feedback configuration that we have seen for the voltage amplifier talking about operational amplifier in Chap. 1, and then, following that analysis, the voltage gain is given by:

$$A_{vf} = \frac{100}{10} = 10$$

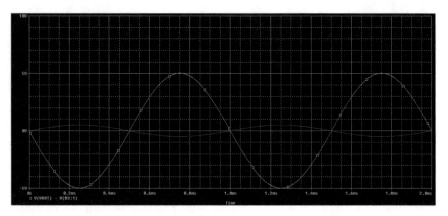

Fig. 3.17 Output (green) versus input (red) signal for the circuit of Fig. 3.15 with 1 kHz sinusoidal signal

To check this point, try to simulate with a different R_4 *and calculate the gain.*

Now, I take the opportunity to talk again about the final stage. Based on what we have said during the positive half-wave Q_{12} is on and Q_{13} off and vice versa during the negative half-wave. This is called 'class B' operation to distinguish it from the 'class A' operation in which the output transistor is on for the whole wave. An example of 'class A' operation is given by a common emitter transistor configuration with a R_L resistor as a load: You have seen thousands of times this simple circuit! If we bias correctly the transistor and put a signal, for example, a sinusoidal waveform, at the input, we will obtain a copy more or less exact of the signal but larger, at the collector of the transistor. The transistor is ON for the whole cycle of the sinusoidal signal, and this is the best situation to obtain an output signal as much as possible similar (low distortion) to the input, but it is most power consuming because the transistor is ON even when the signal is zero. On the contrary in class B, when the signal is zero, there is no power consumption, and each transistor must sustain only half of the total voltage amplitude. However, class B amplifiers are affected by a higher total distortion than class A, this is caused by the point in which the sinusoidal waveform is crossing zero voltage level, and both the transistors are OFF at the same time. This is the reason why some high-fidelity high-power system makers preferred to use a class A configuration for their amplifiers, but then, they had to equip their system with a quite big heat sink to avoid burning everything! Anyway, a third way does exist, and it is to bias both the transistor slightly on at zero signal, so that transition from one transistor to the other can take place smoothly and distortion is minimized. So let's go back to the bias network of final stage: D_1–D_2 sustain $2V_{be}$ between base of Q_{12}–Q_{13}, exactly what is needed to keep them on in DC conditions. The constant current flowing through the diodes, fixed by Q_{17}, is proportional to the bias current of Q_{12}–Q_{13}.

We can use the simulator to check what we have said. We can tie together the base terminals of output transistors just by shorting D_1 and D_2. Now, we have a true class

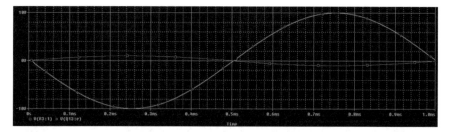

Fig. 3.18 Behavior of amplifier of Fig. 3.19 showing the distortion of the output waveform

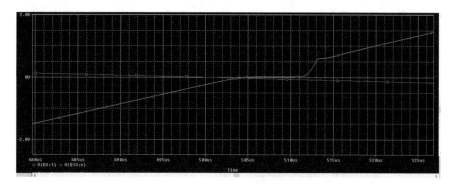

Fig. 3.19 Magnification of the zero-crossing point of output signal remarking the distortion of the waveform

B output stage, and the output waveform we can obtain is shown in Figs. 3.18 and 3.19.

The distortion around zero is clearly visible.

This kind of distortion is called *crossover distortion*, and obviously, it is a characteristic of class B amplifier. The amplifier using a biasing technique of final devices is called a *Class AB* amplifier to underline that in spite of the fact that the operation is mainly that of a class B scheme both the final transistors are in a small conduction if the signal is zero, like in a class A configuration. This is by far the mainstream configuration for power linear amplifiers, for example, for audio applications where a good distortion performance is mandatory.

No load has been applied to the output so far. Try to apply different resistors as a load and discuss the results. What happens if the load is not purely resistive?

Now, we can use simulation to explore the behavior of the amplifier in frequency. It is possible to obtain directly a Bode Plot reporting voltage gain (in dB) and phase (in degrees), that are visible in Fig. 3.20.

The ringing peak of the gain at high frequency signals that the system is unstable in spite of the compensation capacitor C_2 already present. So, we can increase the value of C_2 to reduce substantially the bandwidth but increasing the stability of the amplifier. If we use, for example, $C_2 = 2$ nF, we have the result of Fig. 3.21.

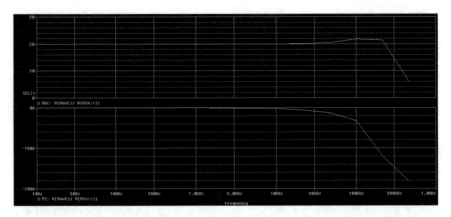

Fig. 3.20 Bode Plot of voltage gain (upper curve) and phase (lower curve) of the amplifier of Fig. 3.15

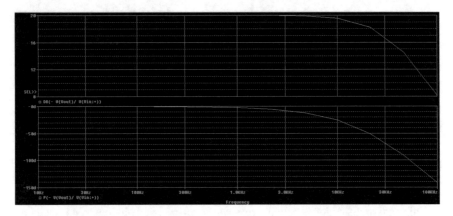

Fig. 3.21 Bode Plot of the amplifier with $C_2 = 2$ nF

In that figure, the upper cut frequency (measured at -3 dB) moved from almost 300 kHz to about 30 kHz, but the benefit in terms of stability is evident: The gain roll-off is smooth and the phase shift is much lower.

So far, we have used a simplified amplifier schematic just as an example. In Fig. 3.22, you can see the schematic of a 'realistic' power audio amplifier. We can discuss the circuit recognizing the scheme of our example, but we will also find something new.

This is an audio amplifier that can drive a loudspeaker of 4 Ω impedance. You can recognize the main blocks used in the previous simplified example, and the input differential stage still has active loads as before, but you can reduce the number of BJT replacing the current mirror with two resistances. Also, the bias current generator of the stage is not a current mirror but a BJT whose base voltage is kept constant by means of diodes. You see at the right of the schematic that the coupling with the

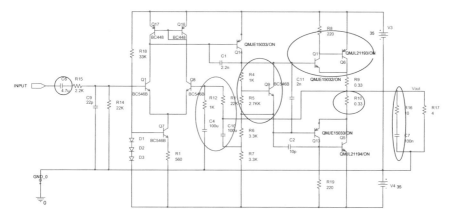

Fig. 3.22 Schematic of audio power amplifier (added relevant component are circled)

input signal is done with a capacitor to avoid the propagation inside the amplifier of an eventual DC component of the input.

Question: What you think are the consequences of this choice on the bandwidth?

The second point is the feedback network that is put on the other input of the differential pair with respect to the input signal. Then, feedback is now node sampling-loop comparison: We know from the theory that this configuration stabilizes the voltage gain and guarantees an high input impedance that can be required by audio transducers. Another important point is that there are two feedback loops, one for the signal that you obtain considering C_4, C_8 as short circuits and another one for DC mode in which output node is directly connected to one input of differential pair through R_{11}. Then, when the input is zero, the feedback loop keeps the output node perfectly to zero. This is required to avoid damaging the loudspeaker that in this scheme is directly coupled to the output. The signal voltage gain is given by:

$$A_{vf} = \frac{R_{11} + R_{12}}{R_{11}} \tag{3.15}$$

Moving to the left, we can recognize the output driver similar to the previous example and the class AB output stage, but you can see that the biasing circuit is now different, no more diodes but the circuit made of Q_9–R_4–R_5. This is a classical configuration in amplifier design called V_{be} *multiplier*. If you look at the schematic, you see immediately that V_{ce} of Q_9 which is also the bias voltage V_{BB} of the output stage that is given by: (neglecting the base current)

$$V_{BB} = \frac{(R_4 + R_5)}{R_5} V_{be} \tag{3.16}$$

So, $V_{BB} = K V_{be}$ which gives the name to this circuit. Generally, R_5 is a trimmer and you can decide in some range the value of K and as a consequence the bias current flowing into the output stage transistors. Generally speaking it has a range of few milliamps, you should keep in mind that an higher current leads to a lower distortion but increases the DC consumption of the amplifier. The strong point of V_{be} multiplier is that if Q_9 is mounted on the same heat sink of the output stage it can track the temperature variations maintain a stable operating point during amplifier operation.

Now, the output stage is made of a complementary Darlington couple to ensure the current capability required from a power amplifier, and you should notice the two small emitter resistances R_9–R_{10}. These are classical feedback emitter resistances which help to minimize the unbalancing of the output BJT due to inevitable parameters spread of discrete devices.

In the circuit of Fig. 3.22, it has been introduced and also the load of the amplifier which is supposed to be a 4 Ω impedance loudspeaker. Even if it has been modeled with a pure resistance, this is not true in the real world and a model of the loudspeaker provided by the manufacturer should be required. To avoid that non-resistive behavior of the loudspeaker that can alter the frequency response, a compensation network made of R_{10}–C_7 has been added to the load.

Well, I think that now, we can try some simulations with our circuit. I report in Figs. 3.23 and 3.24 the Bode Plots and the voltage gain versus frequency.

Bandwidth of the circuit is about 3 Hz–200 kHz. Center band voltage gain is 21, quite in agreement with (3.15).

An interesting simulation is the analysis of the circuit at clipping point. You can see the results in Fig. 3.25.

You can now take your time to make trials changing the key components of the circuit that we have highlighted in this description. For example, it can be changed the trimming resistance R_5 checking the effect on the behavior of the final power stage. In the following, some arguments can be further analyzed:

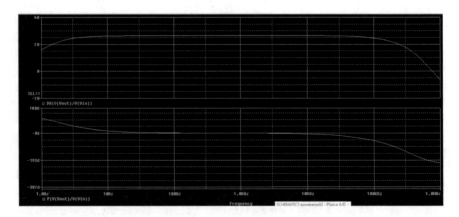

Fig. 3.23 Bode plot of frequency response and phase of the circuit of Fig. 3.22

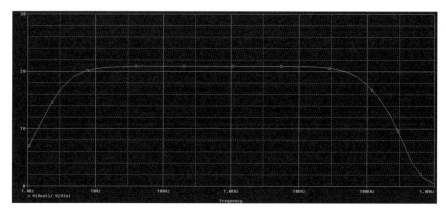

Fig. 3.24 Voltage gain versus frequency

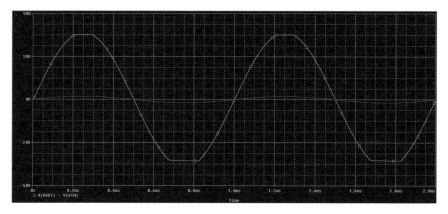

Fig. 3.25 Behavior of the amplifier at clipping point. (output voltage in green, input voltage in red) $f = 1$ kHz

- *Which component controls the roll-off at high frequency?*
- *How can you obtain a smooth roll-off at high frequency?*
- *Can you explain the behavior of frequency response at low frequencies? (0–10 Hz)*
- *Discuss the result of simulation at clipping point (see Fig. 3.25).*
- *Replace the current mirror load in the differential pair with two resistors and discuss the results.*

Chapter 4
Sound Equalizer Using Operational Amplifiers

4.1 Introduction

The analysis of circuits performs sound equalization like it is used, for example, in a hi-fi sound amplifier. It is also an opportunity to talk about processing (filtering) an analog signal.

4.2 Fourier Wave Development Understanding, Square Wave like Infinite Sinusoidal Waves

We suppose you already have a good knowledge about Fourier analysis. You know that every periodical signal, and with some mathematical hypothesis also non-periodical signal, could be represented as an infinite sum of sinusoidal (o cosinusoidal) waves with different frequency and amplitude.

This is quite a great result that helps a lot the analysis not only in electronics but also in mechanics and in all the other technical fields.

A system, not just the electronic ones, could be studied and designed separating the behavior of the system for the different frequency of the signal. When I was a student and I studied Fourier analysis, the professor suggested to do, at least one time in the life, a manual verification: verify that a square wave could be approximated as close as we want, with a summation of sinusoidal waves. So, I designed the sinusoidal components of the square wave, and I did the summation, manually, because the computer was not available at that time.

Today, we can do the same thing but using our virtual laboratory tool. (duty cycle is 50%)

Square wave is defined as:

$$y(t) = -a, \ -\frac{T}{2} \le t < 0 \quad y(t) = a, \ 0 \le t < \frac{T}{2}$$

© Springer Nature Switzerland AG 2020
R. Gastaldi and G. Campardo, *Electronic Experiences in a Virtual Lab*,
https://doi.org/10.1007/978-3-030-45179-0_4

with a period equal to T.

The Fourier analysis gives the sinusoidal series:

$$y(t) = a \sum_{i=0}^{\infty} \frac{4}{(2i+1)\pi} \sin\left(\frac{2\pi(2i+1)}{T}t\right) \tag{4.1}$$

If we use $a = 1$ as the amplitude and $T = 1$ as the period, the first five harmonics are the functions:

$$\frac{4\sin(2\pi t)}{\pi}, \frac{4\sin(6\pi t)}{3\pi}, \frac{4\sin(10\pi t)}{5\pi}, \frac{4\sin(14\pi t)}{7\pi}, \frac{4\sin(18\pi t)}{9\pi}$$

Their amplitudes decrease with respect to the fundamental harmonic as the frequency increase.

Figure 4.1 shows five voltage generators connected with a resistor as a load. Each generator's frequency increases like the Fourier series claims, and the relative amplitudes are calculated with respect to the generator V_1 considered as the principal harmonic.

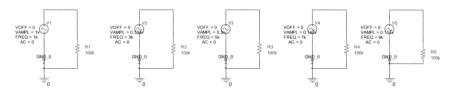

Fig. 4.1 Schematic with five generators with different amplitude and frequency to simulate the first five harmonics for a square wave of 1 V amplitude and 1 kHz frequency

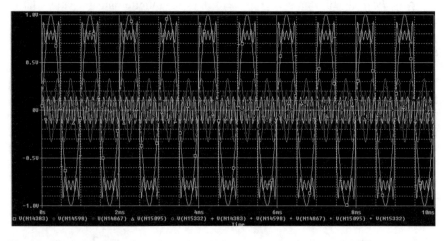

Fig. 4.2 Simulation shows the voltage sinusoidal outputs and their sum in light blue

Figure 4.2 shows the results of the simulation. The first five signals are the output of the voltage sources while the last signal, in light blue, is an addition of the previous five. You can see how the sum can approach a square wave.

My suggestion is to increase the simulation adding other generators, with the right frequencies and amplitudes, like Fourier series requires: twenty generators are normally enough to obtain a good approximation of the square wave.

4.3 Operational Amplifier

The operational amplifier is one the fundamental blocks used in analog electronics. We can say that in the operational amplifier, we can find summarized most of the concept developed in the electronic field circuit arena.

The question is: how people invented this component? What is the thought behind this great solution?

The story starts in the early twenties of the last century. Electronics was done by the vacuum tubes, and the first use was on the telephony, not cellular of course☺!

Harold Black changed the basis of applied electronics by proposing the negative feedback amplifier. This discovery is considered the most important breakthrough of the twentieth century in the field of electronics, since it has a wide area of application. This is because all electronic devices are inherently nonlinear, but they can be made substantially linear with the application of negative feedback. Negative feedback works by sacrificing gain for higher linearity (or in other words, smaller distortion/intermodulation). By sacrificing gain, it also has an additional effect of increasing the bandwidth of the amplifier. However, a negative feedback amplifier can be unstable such that it may oscillate. Once the stability problem is solved, the negative feedback amplifier is extremely useful in the field of electronics.

So, Black proposed the theory, you can find the same drawing he did for his famous patent in many of the electronic book, explaining the advantage of the feedback.

Using the feedback concept, the main electrical parameters for the circuit, gain, frequency bandwidth, input and output impedance depend from the passive components used to build the feedback loop. The feedback amplifier is based on the amplification of the difference of two voltages, one of them is the input voltage and the other is the feedback voltage.

To satisfy this condition, we need to have a component, the operational amplifier, with a very precise characteristic, a relationship between the input and output like

$$v_{\text{out}} = A\left(v_{\text{in}}^+ - v_{\text{in}}^-\right)$$

The output voltage, v_{out}, is equal to the difference of the two input voltage multiplied by a factor A.

A must be infinite (none engineer could use this word, anyway, we are speaking about *ideal* component), input impedance must be infinite, no current can 'enter' inside the input pins, output impedance must be 0, an ideal generator must work on the output and also the bandwidth must be infinite.

One time a friend talked to me about how to explain the bandwidth concept. We were sitting on the chairs and he said: I will clap my hands and, every time I clap you will move your body forward.

He started slowly and I moved my body then he increased the speed to clap and I tried to follow him. After few time, with the increasing of the clap frequency I was not more able to move my body with the required frequency and I stopped myself. He said: 'this is your frequency cut.'

4.4 Voltage Follower

Let's start with a simple circuit, using one of the most famous Op Amp in the industry, the uA741; we have it in the library.

Figure 4.3 shows a voltage follower schematic circuit. To do the schematic, I read the datasheet component, you can find it on the Web, to understand the polarity of the supplies voltages, Fig. 4.4.

I decided to leave the offset compensation pins floating.

The result of the simulation is reported in Fig. 4.5.

From the simulation, you can see that the two signals, input and output, during the ramp are superimposed, and they are different on the top of the simulation, when the input value is 10 V.

Zoom your simulation to check if the two signals have really the same value. In case if this is not true (and this is the case), try to understand why and think how to solve the problem. The same is for the stable part of the simulation.

Ideally speaking, the two input, in this configuration, should have the same voltage value. Please take a moment to understand why. Got it? We are working with an Op

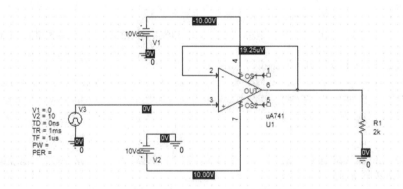

Fig. 4.3 uA741 Op Amp in a voltage follower schematic

uA741

SLOS094G –NOVEMBER 1970–REVISED JANUARY 2018

www.ti.com

5 Pin Configurations and Functions

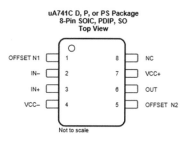

NC- no internal connection

Fig. 4.4 uA741 Op Amp pins configuration from the datasheet

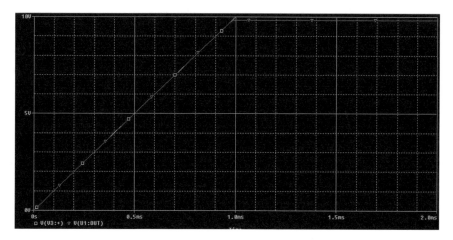

Fig. 4.5 uA741 connected as a voltage follower

Amp with infinite gain, then a finite output voltage can be obtained with a virtually zero differential voltage at the input, which means $V_2 = V_3$. You should realize that in spite of the fact that these two points are virtually connected together no current can flow from one to the other! In fact, the input impedance of the Op Amp is ideally infinite. This is the so-called virtual ground concept, and with this type of concept, we can build a lot of circuit, exploring the real potential of the feedback approach.

Figure 4.6 shows an ideal model for an Op Amp valid for low-frequency operation.

The input connection is realized by a differential pair, analyzed in the previous chapter. R_L is the circuit load.

Figure 4.7 shows the same circuit with resistance series R_1 and R_2 to have the inverting amplifier. We can apply the virtual ground concept to easily understand

Fig. 4.6 Low-frequency
circuit model for an Op Amp

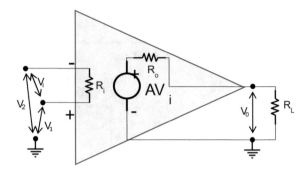

Fig. 4.7 Inverting
operational amplifier
connection

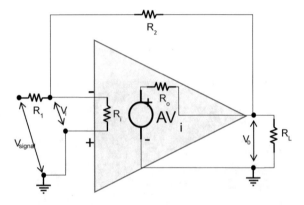

the circuit operation. The $V+$ (V_1) connected to ground has the effect to apply on
the node $V-$ (V_2) the same voltage, in this case ground, obtaining a virtual ground.
$V-$ will be at 0 voltage, but it doesn't sink current because the input resistance R_i is
infinite. In a CMOS differential input stage, the input resistance has the value of a
MOS input gate, really very high, 10^{12} Ω.

This could be possible because the differential input structure, when connected in
feedback with the output, tries to maintain the two inputs to the same voltage value.
Of course, if the input signal changes, we need time to propagate the information
through the feedback and adjust the $V-$ voltage value. This time is directly connected
to the maximum working frequency for the selected Op Amp.

If $V-$ (V_2) is 0 voltage like $V+$ (V_1) is, the network for the current coming from
the signal is 'only' composed by R_1 and R_2. Current from signal flows through R_1
and goes on R_2 before to reach the load.

The Kirchhoff law says:

$$A_v = \frac{V_0}{V_{signal}} = -\frac{R_2}{R_1}$$

So, thanks to the infinite input impedance, the zero output impedance and the infinite gain, we have obtained a virtual ground, and this give to us the possibility to have an amplifier with the gain independent from the Op Amp type and only dependent from the external components, R_1 and R_2 in this case. Really, for me, a very incredible solution!

When I started to study electronics, the virtual ground concept was, for me, one of the most tough concepts, I hope for the reader could be now easier.

4.5 Headphone Non-inverting Amplifier

Let me show another basic circuit that I built for my father many years ago to listen the TV with headphones, a simple non-inverting amplifier. See in Fig. 4.8, the schematic.

Of course, this is one of the many possible circuits that we can use to realize an audio amplifier.

The input signal V_2 is divided by a voltage divider R_{10}, R_{11}; a 20 kΩ potentiometer will be used to control the volume control (R_{11}), however, as this varies. We put also a 20 Ω resistor (R_{10}) to have, in any case, a limit to the input impedance. The signal will flow through C_1 to reach the inverting input (pin3 for the U1A, LF353/NS). The voltage divider R_3, R_4 shifts the audio signal between V_{cc} and ground, and this will also prevent clipping. This is a biasing, putting the signal in the middle of the possible dynamic, 9 V in this case.

At this point, we have a non-inverting feedback with the network done by R_{13}, R_{12} and C_5. Finally, C_3 and R_8 is the decoupling network to separate the DC contribution

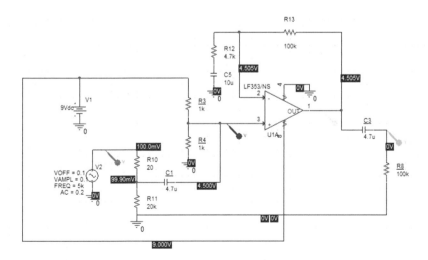

Fig. 4.8 Headphone audio amplifier

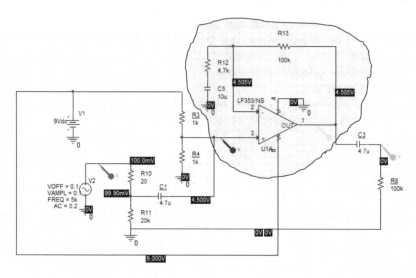

Fig. 4.9 Headphone audio amplifier 'core' circuit

from the signal and send it to the headphone, connected where we have the green probe.

Considering only the net in the blue part as shown in Fig. 4.9, this is the 'core' of the amplifier. The transfer function is:

$$\frac{v_1}{v_3} = 1 + \frac{R_{13}}{R_{12} + \frac{1}{sC_5}}$$

With a *zero* in the origin and a *pole* at ~0.3 Hz.

Let's see first the simulation done with a TRAN option, in the time domain, as shown in Fig. 4.10.

In the simulation setting, we can choose the AC sweep analysis as shown in Fig. 4.11.

The result is showed in Fig. 4.12.

In Fig. 4.12, you can see the value, in dB, of the gain versus the frequency.

We have to take in account not only the feedback network but all the components. The results show a flat curve from 100 Hz to more than 20 kHz, as required by the audio project. The gain value is in the flat zone:

$$\text{Gain} \sim \frac{R_{13}}{R_{12}} = 21.27$$

From Fig. 4.12, the gain, in dB, is around 26.5, so that the gain value is:

$$\text{gain} = 10^{\frac{26.5}{20}} \sim 21 \tag{4.2}$$

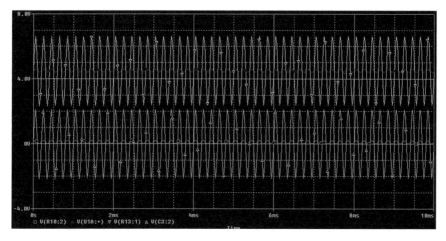

Fig. 4.10 Headphone audio amplifier time simulation

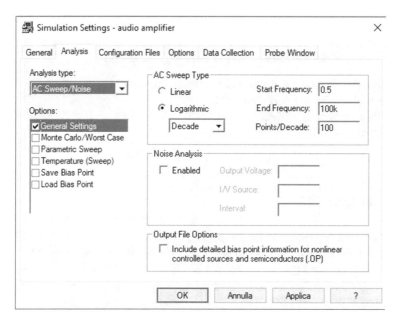

Fig. 4.11 Frequency-domain analysis setting

I hope you found it as a big 'mistake' in the circuit. It is not a mistake but a not true value. In the schematic as shown in Fig. 4.8, the load is 100 KΩ. If this represents the impedance of the loudspeaker is not true. Normally, the impedance is 8–16 Ω. The fact is that I did the circuit, many year ago, using another operational amplifier able to deliver more current but, this Op Amp is obsolete and the model doesn't exist so, the concept is right but the realization will fail. You can try to substitute the

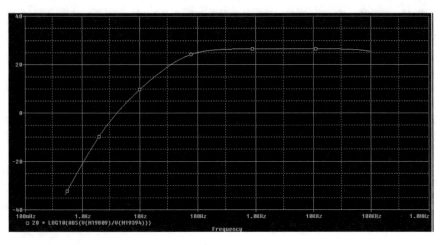

Fig. 4.12 Frequency-domain analysis

operational or putting a BJT to be driven by the Op Amp output, which is able to drive an 8 Ω load.

The proposal, for the reader is to try to modify the circuit:

- *change the value of the temperature in the simulation*
- *change the value of the components*
  Change the bias R₃ and R₄.
  Change or remove C₃.

4.6   Bandpass Filter

Let's do something about the bandpass filter.

Figure 4.13 shows a bandpass filter, a parallel resonant configuration. The output, where it is positioned the red probe, is a result of the dividing impedance between the input resistance R_7 and the parallel between R_8, L_2 and C_2. Of course, the parallel value is dependent from the frequency value and so the same for the output. Figure 4.14 shows the related frequency analysis.

We can see a peak on the simulation, around 160 Hz. On this frequency value, the output, the red probe voltage, has the maximum value. For the resistor divider means that, at that frequency, the total impedance of the parallel circuit will have the maximum value in ohm.

The proposal, for the reader is to calculate:

- *using Laplace formulas, the Bode diagram and the zero and poles values*

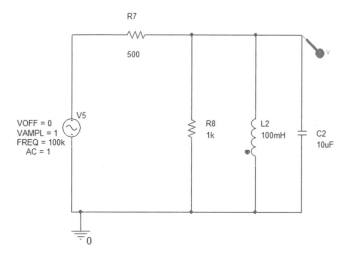

Fig. 4.13 A bandpass filter

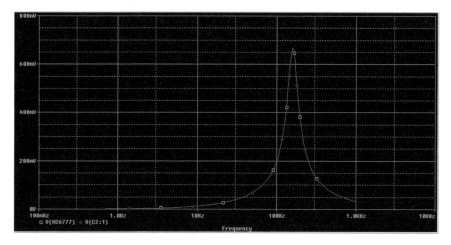

Fig. 4.14 A bandpass filter frequency analysis

This type of circuital configuration could be used to select a frequency in a signal composed by different frequency values, like, for example, an antenna receiver.

With a different combination, we can provide circuit to stop only some frequency, the reverse of the bandpass circuit.

You have to imagine this to be an electromagnetic wave travelling in our circuit; at low frequency, the inductance L_2 is a short circuit; and the voltage on this terminal, the output, will be very low, vice versa; at high frequency, the capacitor C_2 will have a very low impedance, shorting the output signal to a low voltage. In the middle of the frequency, there will be a continuum with a maximum.

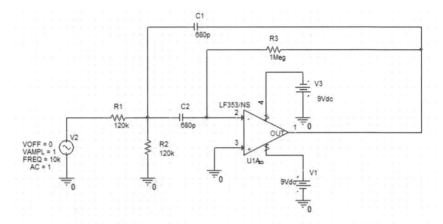

Fig. 4.15 A bandpass filter schematic

How to build a band stop filter?

Using passive component, it is possible to manage the signal frequency composition, but the results, the output, will be always a signal lower, in voltage value, with respect to the input. What would be really useful is to have a filter able to select and amplify: we can do it using an operational amplifier.

To perform a parametrical analysis, we can consider the schematic in Fig. 4.15 for getting, for this moment, the values of the components. We would like to write equation for the gain, quality factor and the filter center frequency. To write the equations, we can calculate head on or to apply some very interesting rules. First, an overview on the schematic: At low frequency, C_2 will be an open circuit so $V-$ input will be at ground, like $V+$; the output will be at ground. At high frequency, C_2 will be a short circuit, as also C_1 is a short circuit; R_3 results in parallel with C_2, and the parallel C_2, R_3 will be a zero impedance, so, again, the output will be at ground.

'Open a window' on the automatic feedback control systems theory: how we can estimate the number of the *poles* and *zeros* in a transfer function only by seeing the schematic?

First, we have to remember that the number of the *poles* is equal to the number of the independent energy storing elements in our network.

Independent means that for the capacitor considered (or inductor), we can assign a voltage independently from the other energy store components present in the network.

In Fig. 4.15, if we assign a voltage for $C1$, one of the terminal of the operational amplifier, $V-$ is at ground, so assigning a value to the other terminal is not enough to assign a value on the output, then we can say that our circuit has 2 *poles*. To know the number of the zeros, in general, we have to know the value of the gain function, and for $s \to \infty$, the number of the *poles* is always greater than the number of zeros. If m is the number of the *poles* for $s \to \infty$, the number of the *zeros* is $m - 1$. For our circuit, we have one *zero*.

Fig. 4.16 A resonant filter

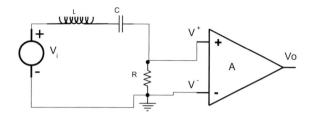

Figure 4.16 shows a circuit that we can use to do some calculation. RLC circuit represents a resonant circuit. Low frequency shows the L like a short circuits but C like an open circuit, so the input will be at ground and so the output. With the frequency increase, L and C become more resistive, and the input will start to assume a positive value, then also the output will have a positive value. When the frequency increases, C becomes a short circuit, and on the contrary, L an open circuit and the input will return at ground while the output voltage will do the same.

The equation for the voltage gain is

$$A(jw) = \frac{V_0}{V_i} = \frac{RA}{R + j(wL - 1/wC)}$$

Center frequency is defined as the frequency for which

$$w_o^2 = \frac{1}{LC} \tag{4.3}$$

The meaning of (4.3) is that at frequency $f_0 \equiv \frac{w_o}{2\pi}$, the reactance of the inductance and the capacitor have the same value.

Quality factor Q and bandwidth are defined by

$$Q = w_o \frac{L}{R} = w_0 \frac{1}{RC} = \frac{1}{R}\sqrt{\frac{1}{LC}}$$

$$B = \frac{f_0}{Q} = \frac{1}{2\pi}\frac{w_0}{Q}$$

Now, we can write the voltage gain equation as:

$$A(s) = \frac{V_0}{V_i} = \frac{\frac{w_0}{Q}As}{s^2 + \frac{w_o}{Q}s + w_0^2} \tag{4.4}$$

This equation is the general expression for the second-order bandpass filter, 2 *poles* and 1 *zero* in the origin. It is possible to have an inductance integrated but the value could not be so big, and rules of thumb can say that we can have 1 nH for each millimeter, too low; so, a bandpass configuration must be obtained using only capacitors, like in Fig. 4.15. A filter is represented by its *poles* and *zeros* number.

Applying Kirchhoff and Ohm's law, we obtain, with $C_1 = C_2 = C$:

$$\frac{V_2 - V_x}{R_1} = sC(V_x - V_{out}) + sCV_x + \frac{V_x}{R_2}$$

V_x is the node voltage connecting R_1, R_2 and Cs. Second equation is:

$$V_x = -\frac{V_{out}}{sCR_3}$$

which is obtained with the hypothesis that current entering the operational amplifier's pin2 is zero.

With some algebraic elaboration, we can write the gain as:

$$\frac{V_{out}}{V_2} = -\frac{\frac{s}{R_1 C}}{s^2 + \frac{2s}{CR_3} + \frac{1}{R'}\frac{1}{C^2 R_3}} \tag{4.5}$$

$$R' = R_1 // R_2$$

All this algebraic job is done to have a formula with the same shape like Eq. (4.4). Equation (4.5) reaches our scope, and we can match and compare (4.5) with (4.4).

$$w_0^2 = \frac{1}{R'}\frac{1}{C^2 R_3} \rightarrow f_0 = \frac{1}{2\pi C}\sqrt{\frac{1}{R_3}\frac{R_1 + R_2}{R_1 R_2}}$$

This is the expression for the center frequency.

$$\frac{w_0}{Q} = \frac{2}{CR_3} \rightarrow Q = \pi f_0 C R_3 \tag{4.6}$$

And the (4.6) gives the expression of the quality factor Q. With the same approach, we can calculate the equation relationship for the resistance:

$$\frac{wA_0}{Q} = \frac{1}{R_1 C} \rightarrow R_1 = \frac{Q}{2\pi f_0 A_0 C}$$

$$\frac{2}{CR_3} = \frac{w_0}{Q} \rightarrow R_3 = \frac{Q}{\pi f_0 C}$$

$$R' = \frac{1}{2Cw_0 Q}$$

With the last expression, we can recover the values of R_2 if necessary.

4.7 Sound Equalizer

Coming to our circuit as shown in Fig. 4.15, we can start to assign the value to have our sound room equalizer.

An equalizer is a circuit able to change the composition of the sound in input by suppressing or highlighting frequency components. This permits to equalize different ambient response; materials can absorb specific frequencies changing our perception of the sound.

A sound equalizer is a set of bandpass filters with a different center frequency.

Input sound, composed by all the frequencies, reaches all the bandpass filters but each frequency will be processed by the specific filter that could exalt or decrease the gain for each frequency, more precisely for each bandwidth. Setting the values:

$$A_0 = 4$$

$$Q = 2$$

$$R_1 = 120\,\text{k}$$

For the input value resistance

$$R_1 = R_2 = 120\,\text{k}$$

$$R_3 = 1\,\text{Meg}$$

$C_1 = C_2 = C$ will be defined depending on the center frequency for the different frequency bandpass filter. We say:

Now, we can start to simulate our circuit. Before simulation, we can find some useful relationship for the most important parameters.

$$A(s) = \frac{V_0}{V_i} = \frac{-\frac{s}{R_1 C}}{s^2 + \left(\frac{2}{R_3 C}\right)s + \frac{1}{R^1 R_3 C^2}}$$

With $C_1 = C_2 = C$, and

$$R^1 = \frac{R_1 R_2}{R_1 + R_2}$$

$$Q = \pi f_o C R_3$$

$$f_0 = \frac{1}{2\pi C}\sqrt{\frac{R_1 + R_2}{R_1 R_2 R_3}}$$

Table 4.1 C capacitor value and relative center frequency

f_0 (Hz)	C
32	0.022 μF
64	0.011 μF
125	0.0056 μF
250	0.0027 μF
500	0.0015 μF
1 k	680 pF
2 k	330 pF
4 k	160 pF
8 k	82 pF
16 k	43 pF

$$R_1 = \frac{Q}{2\pi f_0 A_0 C}$$

$$R_2 = \frac{Q}{2\pi f_0 C (2Q^2 - A)}$$

$$R_3 = \frac{Q}{\pi f_0 C}$$

With the help of the equations, we can start to simulate.

We understood that we need to have one bandpass filter for each frequency band. Total band frequency, for an audio project, is between few tens of Hz to 16 kHz. We decided, by project, to have ten bandpass filters, each of them centered on different frequency like in Table 4.1.

We have to design ten filters, with different values of the capacitors, with values showed in table.

The total signal must arrive from one input and given to all the bandpass; all the bandpass filters must have the inputs shorted.

The outputs of the different filter will be added again through a potentiometer, one for each filter changing the value of the gain. The problem is that: if our input signal has 1 V maximum value of amplitude and we send the signal to ten different filters that select frequency but not change amplitude, we can have, like filters output, ten signal by 1 V for a 10 V maximum amplitude sum.

To manage this situation, we will use an input amplifier with a 0.25 gain value while, as we already said, the gain for the bandpass filter is four, then the sum of all the filters' output will be done with a potentiometers designed to have in the center of them, a total gain by unit.

Figure 4.17 shows the input amplifier, UA2, with the network R_7, R_8 to a gain equal to 0.25. The resistor divider R_8, R_9 do not change the value of the input because node $V-$ of UA2 is a virtual ground. The function of R_9 is only to stabilize the OP AMP. So, the gain of the input amplifier is $R_7/R_8 \sim 0.25$.

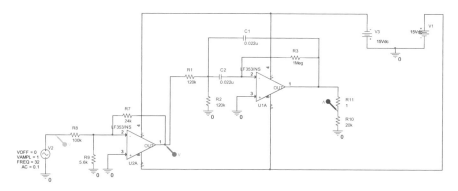

Fig. 4.17 A resonant filter with amplifier in input

Figure 4.18 shows the AC analysis showing the working mode in frequency of the circuit.

Resistor divider R_{11}, R_{10} represents the potentiometer (the potentiometer is a 20 kΩ). This simulation is at its maximum value.

Figure 4.19 is the complete schematic for the room equalizer. We designed only two bandpass filter, the first one, centered on 32 Hz and the last on 16 kHz.

The potentiometer is put in the middle of their range, 10 kΩ, and we connect also the summing amplifier U2B. Gain of the adder is ~4, but with the capacitor C_5, we also add a *pole*. *Why? What happened when the global input signal will be zero?*

Design the complete room equalizer with all the ten bandpass filters.

Study and simulate the variability of the design with the temperature.

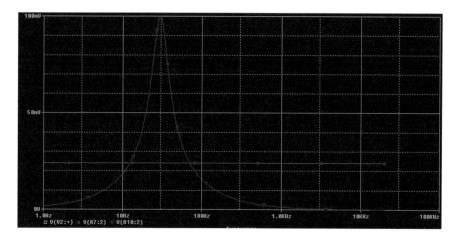

Fig. 4.18 AC analysis for circuit on Fig. 4.17

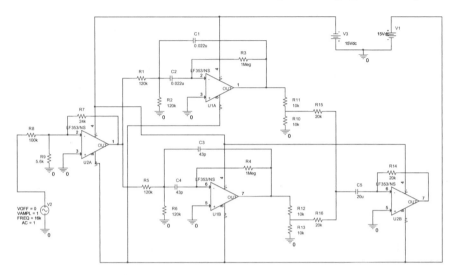

Fig. 4.19 Schematic of an audio room equalizer

Chapter 5
Digital Circuits

5.1 Introduction

In this chapter, I would like to make some experiments with digital circuits: yes, the stuff including logic gates, flip-flops, decoders, registers, memories and other circuits of increasing complexity up to microprocessors that are the heart of most of the electronic systems that populate our daily life.

Our experiments will not deal with microprocessors of course but with something really more simple: the first circuits we will look at are counters. The reason why I choose this matter is that counters have many applications in electronic systems we use every day and they are a simple example of sequential circuits which means we can have an idea of a system that evolves, i.e., walks through different states depending on external inputs and the *past history*. A counter then is a *state machine* which is a key concept to understand automata and processors operation. We will use counter concepts to build an electronic clock.

Later we will look at arithmetical elements: they are a very important class of logical circuits, as the main function of most computers is to perform arithmetic operations, so they need arithmetical elements. We will see the behavior of a full adder.

5.2 A Simple Decade Counter

First of all, we need a flip-flop (FF). You certainly know this circuit from basic courses in electronics, and you also know there are many types of FF, each one characterized by a truth table. The truth table is a compact but complete description of the operation of each logic circuit: it describes the logic level of the output(s) for each possible logic level present on the input(s) and it is all what you need to use the circuit from a logic viewpoint. In the case of FF, the state of the output at time t depends not only from the state of the input at that time but also from the state of the output at

© Springer Nature Switzerland AG 2020
R. Gastaldi and G. Campardo, *Electronic Experiences in a Virtual Lab*,
https://doi.org/10.1007/978-3-030-45179-0_5

Table 5.1 Truth table of a *JK*-FF

CK	J	K	Q_{i+1}	Q^*_{i+1}
↑	0	0	Q_i	Q^*_i
↑	1	0	1	0
↑	0	1	0	1
↑	1	1	Q^*_i	Q_i

the time $t - 1$. But the time in a circuit like a FF, which is called *sequential* is not a continuous variable as we are used to consider in everyday life: actually, something happens in the circuit only at the instants in which the clock makes a transition, (low to high or the contrary depending on the design of the circuit) then we could say in some way that clock edge operates a discretization of the time into single instants,... $t - 1, t, t + 1$... and so on.

The FF we will use in our experiment is the well-known *JK*-FF. Let's give a look to his truth table in Table 5.1.[1]

In this table, we can see that every output change is triggered by the rising edge of the clock (but, I repeat, we could also design the circuit to use the falling edge). In the most right columns, the status of the output Q at the time $i + 1$ is reported with respect to the status at the time i. For our counter, we are interested in the last row which reports that if $JK = 1$, output Q_{i+1} is always the complement of Q_i.

We can then draw the schematic of a four bit counter in Fig. 5.1.

Before starting to talk about the schematic of Fig. 5.1, it is necessary to say few words about simulation of digital circuits using ORCad Pspice. The simulator SW gives the possibility to simulate a circuit in which analog and digital components coexist (mixed-mode simulation). Modeling of digital components is done having in mind the *logic* behavior of the device. For example, in Fig. 5.1, I have used a model

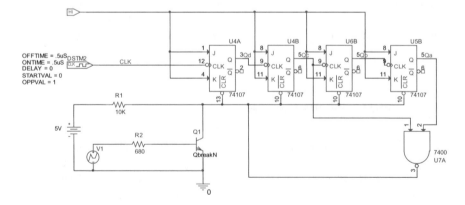

Fig. 5.1 Schematic of a 4-bit binary counter

---

[1] Through the text the character "*" after a logic variable indicates the complement of that variable, then, for example, Q^*_i means the complement of Q_i.

of *JK*-FF available in the simulator library. Inputs and outputs of this model are considered as logic signals, so they are characterized by logic levels 'high' and 'low' but the details of the waveform in the time domain are not analyzed except for the delay between input and output which is a parameter assigned to that component. The simulator anyway accepts a mix of analog modeled and digital modeled components and can analyze a circuit containing both of them. In Fig. 5.1, for example, the pulse generator V_1, resistors R_1 and R_2, the transistor Q_1, and 5 V voltage source are all analog components, then the signal entering pin10 (CLR*) of all the FFs will have a time-dominium detail like a physical signal while the output Qs of the same components will be a pure logic signals having then only high and low states and eventually a delay with respect to CLK or other inputs like J or K. I don't want to go on in the discussion about mixed-mode simulation because you can find a more detailed description in Chap. 9. Let's come back to our counter now. Please give again a look to Fig. 5.1. You see four *JK*-FF, a clock generator; a NAND gate and few other components. Why I need four *JK*-FF? The answer is quite simple: we want to count a decade but digital circuits work with binary numbers and each FF represents a digit or better a bit of a binary number. The question now is: how many digits I need to obtain the decimal number 10? Obviously, four because n digits will allow 2^n combinations, so 4 digits allow 16 different numbers, one of which can be associated with the decimal number 10. If I had only 3 FFs I would have been able to do only eight combinations ($2^3 = 8$), not enough to count a decade.

The four FF are connected in cascade the output Q of the first feeding the CLK input of the second and so on for all the components. CLR* inputs are connected together to an inverter stage made with the transistor Q_1. A pulse generated by the voltage source V_1 resets all the output to zero at start-up.

If we now connect our clock generator to input CLK of the FF named U_{4A}, the circuit will start to complement Q output of each FF at the high-to-low transition of incoming pulses exactly as stated by Table 5.1 except for the sign of clock transition until all the sixteen 4-bit combinations have been made and then restart the never-ending cycle. Q output for every *JK*-FF of the circuit drawn in Fig. 5.1 can be collected in Table 5.2.

It is easy to associate each *JK*-FF configuration in Table 5.2 to a 4-bit binary number corresponding to the decimal number in the first column and considering Q_A as the most significant bit and Q_D as the least significant bit. As $2^4 = 16$, we will have sixteen possible numbers. The transition from one number to the next is triggered by the falling edge of the clock CLK. Now, my problem is that I want to build a counter by 10 and not by 16, then I need something that is able to stop the count after ten pulses and reset the system to zero. You can imagine that the NAND gate designed in the schematic is the solution to my problem, but let's see how it works.

We have to come to Table 5.2, we can see there that every number is obviously represented by a unique combination of Q outputs, for example the number 10 is characterized by the presence of a "1" on Q_a and Q_c, it is the only number to have this combination, then if we take a 2-input NAND gate and connect an input to Q_a and the other to Q_c while the output is connected to the terminal CLR* of each FF we should

Table 5.2 Q output of the four JK-FF in the circuit of Fig. 5.1

N_o	Q_a	Q_b	Q_c	Q_d
Reset	0	0	0	0
1	0	0	0	1
2	0	0	1	0
3	0	0	1	1
4	0	1	0	0
5	0	1	0	1
6	0	1	1	0
7	0	1	1	1
8	0	0	0	0
9	1	0	0	1
10	1	0	1	0
11	1	0	1	1
12	1	1	0	0
13	1	1	0	1
14	1	1	1	0
15	1	1	1	1

Transition to the next configuration happens at the falling edge of CK

reset our counting just after ten CLK pulses. In this circuit, we have introduced a decoder, or in other words a combinatory network made of the 2-input NAND gates that produces a '0' on the output when Q_A and Q_D are at '1' at the same time. Let's now start the simulation and have a look at the result in Fig. 5.2

If you look at the upper part of the figure, you can see the four Q outputs of FFs as digital signals.

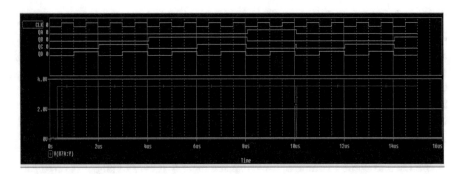

Fig. 5.2 Operation of a 4-bit binary counter

In the lower part of Fig. 5.2, it is drawn CLR*. This waveform is analog, and then the voltage level and its real shape are reported. The simulation program decides automatically which signals must be considered analog and which must be logic.

If you sweep the red cursor from left to the right, you can see the combinations coming out at the low transition of clock pulses, exactly in the order predicted in Table 5.2. At the time specified by the glitch of CLR* the sequence ends and will then restart from the beginning. A detail that it is interesting to note is that in this circuit configuration the input-to-output delay t_d of one FF is transferred to the next so that if we have n FFs the total delay in case all the FFs make a transition is equal to nt_d. This is shown in Fig. 5.3 where all the outputs make a transition moving from 0111 to 1000. You can see very well the cumulated delay to reach the final configuration with respect to the clock transition. Of course, this problem is caused by the fact that each FF changes his state in an asynchronous way because the CLK signal is coming from the previous FF in the chain and this may cause problems when real delays of the components are taken into account.

Another problem can be highlighted looking at Fig. 5.4 which is an enlargement of the point in which the counter is resetted, having counted ten pulses.

If you look at Fig. 5.4 you can see that the configuration 1010 at which the counter resets is actually no more than a glitch.

This is not a malfunction but the logic behavior of the circuit: infact, as soon as the configuration 1010 appears the signal CLR* is immediately generated leading to the reset of all FFs, anyway it may be a problem to use this configuration to drive some other circuit, for example, a decoder just because it is present for a short time.

You have already understood that this counter is an asynchronous circuit, or in other words, the switching of the FFs happens at different times because the CLK

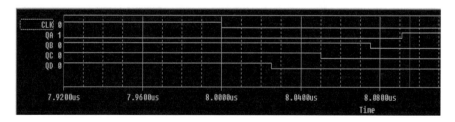

Fig. 5.3 Detail of the switching delay of the FFs with respect to clock during the transition 0111 to 1000

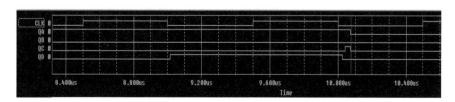

Fig. 5.4 Decade counter reset point

pulse propagates through the FF's chain and each switching instant depends on the delay of a FF.

I think that now it's the time to ask you some questions to play with your counter circuit:

– *How we can model the delay introduced by the connections of the circuit? For example, how we can take into account the fact that the CLR* pulse is not applied at the same time to all the FFs? And what happens in this case?*
– *If we introduce a model for taking into account the delays of connections and of FF's what happens if we increase the CLK frequency?*

Another way to build a counter is to feed the FFs with CLK in parallel so that switching time is approximately the same for all of them: we call such a circuit a *synchronous counter*.

I would like to talk a bit about it because it is a good example of how a *state machine* works.

First of all, let's see a block diagram of a synchronous counter (see Fig. 5.5).

You see from this figure that it is a particular state machine in which there are not external inputs except the CLK pulse and outputs are the state variables. The design of this counter is more complicated with respect to the previous one because we have to design the combinatory network. Let's go on.

First of all, we have to decide all the states through which our counter must go. Let's suppose to realize again a decade counter, then we can draw a diagram like in Fig. 5.6.

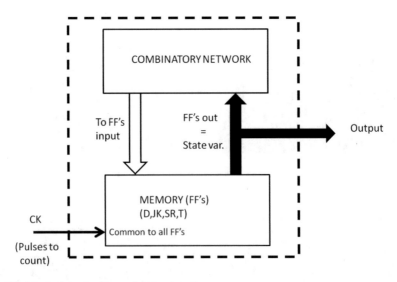

Fig. 5.5 Block diagram of a synchronous counter

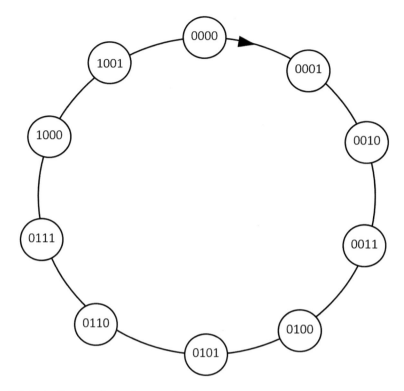

Fig. 5.6 State diagram of a synchronous decade counter

Starting from the state 0000, the counter will move clockwise at each clock pulse turning back to 0000 after ten pulses. Please note that we are not obliged to choose this particular sequence for our states but we can choose a different one.

Little question: *can we do the same with an asynchronous counter?*

Did you come to an answer? Good, let's go on. Now, we have to translate these states into a table in such a way to be able to design an appropriate combinatory network. I have done it for you but before you look at the Table 5.5. I'll tell you that I have used *T*-FFs instead of *JK*-FFs as memory elements. To keep the *JK*-FFs of previous circuit in the future simulation, it is enough to short together *J* and *K* input as you probably know. Well, please have a look to Tables 5.3 and 5.4 which contain all the information we need to design the combinatory network.

Table 5.3 Transition table of *T*-FF

T_n	Q_{n+1}
0	Q_n
1	Q_n^*

Table 5.4 Excitation table of the counter

Q_3	Q_2	Q_1	Q_0	Q'_3	Q'_2	Q'_1	Q'_0	T_3	T_2	T_1	T_0
0	0	0	0	0	0	0	1	0	0	0	1
0	0	0	1	0	0	1	0	0	0	1	1
0	0	1	0		0	1	1	0	0	0	1
0	0	1	1	0	1	0	0	0	1	1	1
0	1	0	0	0	1	0	1	0	0	0	1
0	1	0	1	0	1	1	0	0	0	1	1
0	1	1	0	0	1	1	1	0	0	0	1
0	1	1	1		0	0	0	1	1	1	1
1	0	0	0	1	0	0	1	0	0	0	1
1	0	0	1	0	0	0	0	1	0	0	1

In Table 5.3, you can see that the FF input T must be 1 only if a state transition is required. This information must be used in Table 5.4, where you can find on the left in blue the output Q_3–Q_0 of FFs at the time t and near them in brown the output Q'_3–Q'_0 at the time $t + 1$. You can recognize that they reflect the walking through the states described in Fig. 5.6. Signal Q_3–Q_0 are then the *state variables and also the outputs* of our state machine. Now, we must determine what is the input we must apply to the FFs ($T_3 - T_0$ signals) at the time t to produce the desired state transition at the time $t + 1$. You find it in the four columns on the right of the table and the two red arrows highlight two examples of how the green part of the table has been filled.

In the uppermost example, we see that Q_3 is 0 at time t and must remain 0 at time $t + 1$, and then looking at Table 5.3, we must put $T_3 = 0$, while in the lowermost example, Q_3 must change, then $T_3 = 1$. We can translate the green part of Table 5.4 in some algebraic equations defining the combinatory network we have to design:

$$T_3 = Q_0 Q_1 Q_2 + Q_3 Q_0 \tag{5.1}$$

$$T_2 = Q_0 Q_1 \tag{5.2}$$

$$T_1 = Q_3^*/Q_0 \tag{5.3}$$

$$T_0 = 1 \tag{5.4}$$

Exercise: *Try to find the equations above starting from* Table 5.4 (*you will need Karnaugh maps*).

We can now draw our decade counter (see Fig. 5.7).

It is easy to recognize that the added logic gates are the translation in circuital terms of the Eqs. (5.1)–(5.4). These gates are the equivalent of the combinatory network of Fig. 5.5 and their scope is to reset the counter after ten clock pulses. Also, a signal CLR is present to correctly set the counter at the start-up. The combinatory network is more complicated than in the asynchronous counter, but in this circuit, the delays of FFs are not added and the output configuration is stable for a full clock period.

Let's see the output of simulation in Fig. 5.8.

Once again the red cursor can be moved from the left to the right to verify the correct sequence of states. You can count the number of CLK pulses (remember to

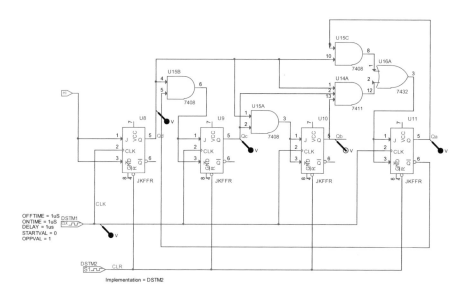

Fig. 5.7 Schematic of a synchronous decade counter

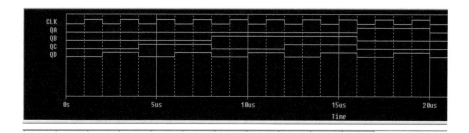

Fig. 5.8 Output waveforms of synchronous decade counter

refer to the falling edge of CLK) and verify that after 10 of them; the counter is back on 0000 state. If you try to zoom in correspondence of the falling edge of the clock, you will find that all the outputs switch in a synchronous mode.

If you want to change the sequence of states, this will result in a different implementation of the combinatory network. Please try it!

Question: In the asynchronous counter, is it possible to freely choose the sequence of states?

5.3   An Electronic Clock

One of the most diffused application of counters is the electronic clock. We are surrounded by electronic clocks: on our wrist, on the car, near our bed and in every electronic device in our house an electronic clock is incorporated. So, I thought this is a good example to show the use of counters in electronic systems.

An electronic clock is made of a reference frequency generator which is able to deliver one pulse per second, a number of counters for seconds, minutes and hours and a number of decoders to send the counter code to some display device.

Let's see the overall clock schematic in Fig. 5.9.

In this figure, we can see that the circuit of the clock has been built using some 'hierarchical blocks' whose content is not visible at this level, that can be connected each other and with other elements of the system through some I/O pins placed at the border of the hierarchical block. This is a very useful method to draw complex circuits in a compact way. You can visualize and operate on the content of each sub-block simply clicking on it. Creating a hierarchical block is very easy using the interactive interface of capture tool: you have to click the related button under 'place' and define the I/O pins, then enter the block and draw the circuit inside, placing the ports corresponding to external pins where you want to connect the internal circuit to the external elements.

To understand the schematic of Fig. 5.9, let me say first of all that it represents only the part of the clock that counts seconds and minutes. So, with this schematic, you can count up to one hour or more exactly 59 min and 59 s. On the next second, this part of the clock will be reset to 00:00 and a pulse (CKout1 in the schematic) will be sent to the section that counts 24 h.

The reason why I made this simplification is that a simulation involving hours is too long to be done even if the reference frequency is increased to shorten the simulation time. Moreover, the section of hours doesn't add anything new for our learning.

At the bottom of the schematic, you can see a sine wave generator with frequency 50 Hz and amplitude 220 V which simulates our home power line, and of course, there is also a voltage transformer TX1 to reduce the voltage applied to the block CKGEN that has many important functions:

– It must provide the DC power supply to all the clock system.

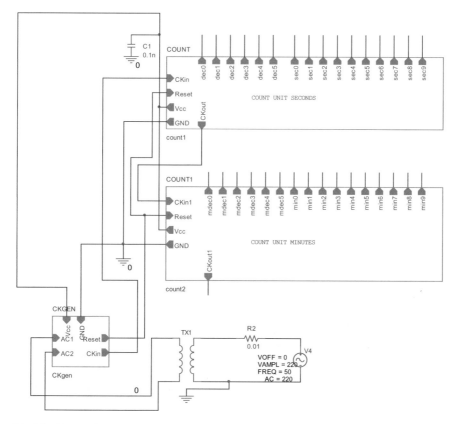

Fig. 5.9 Circuit of the electronic clock (only the section of seconds and minutes has been included)

- It must generate the reference frequency for the clock.
- It must provide a reset pulse for all the counter units at the start-up of the system.

Let's go through these functions in detail looking into the block CKGEN (see Fig. 5.10).

In the schematic of Fig. 5.10, I have highlighted in red the section dedicated to the generation of DC power for all the clock The AC voltage is rectified by a diode bridge and sent to a series regulator made of transistor Q_2 and Zener diode D_2 which set the base voltage of Q_2 to 5.1 V, and then the emitter will stay at 5.1–0.6 V, about 4.5 V. The current needed by the circuit is supplied by Q_2 which adapts its V_{CE} to keep stabilized the output voltage of 4.5 V. C_1 and C_2 are simply filter capacitors to reduce the ripple of AC power line. You can see the result of simulation in Fig. 5.11.

You can see in red the output of diode bridge rectifier, about a sawtooth waveform with a half time period with respect to the input sine wave. The blue line is the regulated output voltage, its ramp-up time is regulated by the filter capacitors C_9 – C_2 and the residual ripple you can see in Fig. 5.11 is negligible. You can also see in green the 50 Hz waveform that will be used to generate the reference pulse for the

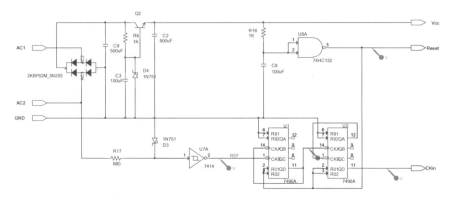

Fig. 5.10 Schematic of block CKGEN (relevant signals probing points are indicated)

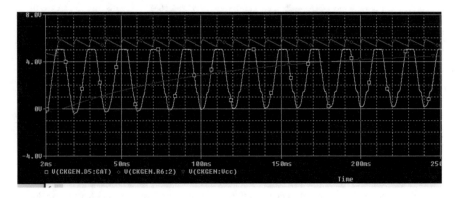

Fig. 5.11 Simulation of voltage regulator operation during ramp-up

clock that is produced by the Schmitt trigger U_{7A}, ensuring a clean square output signal that can be used as a reference for our clock.(signal named REF).

You may think that the home power line is not a precise frequency reference; on the contrary in spite of the fact that a quartz oscillator is of course a more precise frequency reference home AC line has a 50 Hz frequency quite precisely controlled, well enough for a home clock, because errors may cause delivered electrical power not to be used efficiently and then a waste of money by users.

In Fig. 5.12, you can see the result of simulation of the circuit that generates one pulse per second. The signal REF has to be divided by fifty, then it is sent to a first divider-by-five. I have used an integrated circuit (IC) 7490A: you can easily find its datasheet into the Web; then, I will not report here the connections to the pins as you can find them into the datasheet itself. I only point out that 7490A is actually made of a divide-by-five and a divide-by-two stage and you can connect them together (see the datasheet) to obtain a divide-by-ten circuit. The output of this IC is sent to a second 7490A used this time in the divide-by-ten configuration. The final result is

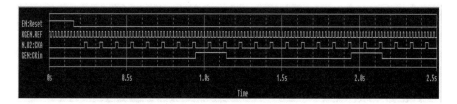

Fig. 5.12 Reference pulse generation inside CKGEN block

a divide-by-50 stage (5×10) producing an output signal named CKin, which is the actual input to the counting stages.

Figure 5.12 also shows the signal CKA which is the intermediate reference obtained by dividing the 50 Hz power supply by five.

The simulation shown in Fig. 5.12 covers more than 2 s and you can see correctly two pulses of CKin, but the simulation has been performed considering also the power-on of the circuit.

From the figure you can see the signal REF (remember that it is the result of squaring the 50 Hz power supply) which needs to be divided by fifty to obtain a pulse per second that the amplitude of the input pulse (purple line) to the divide-by-50 stage follows the ramp-up of the internal supply voltage provided by the regulator circuit. You can also see that the reset signal (red line) keeps all the counter of the system blocked until the internal supply voltage has reached its steady-state value, then counting starts after reset signal has gone low. This is very important to avoid malfunction during the count operation.

The last circuit we need to analyze to finish the analysis of the CKGEN block is highlighted by a blue line and must generate a reset signal at the start-up of the clock, or in other words when we switch on the power to the system. This is necessary to avoid an incorrect positioning of all the counters. Such a circuit is very common in logic design and it is called *power-on reset*. It is apparently very simple but it may be the source of a number of failures because its tuning is very critical. Actually, it must generate a signal only when the power supply has reached its steady-state value, not too early otherwise the power supply will be too low and still moving and more important reset signal must be generated in a reliable way, which means *always when it is needed* regardless tolerance of electronic components, operation temperature and power supply variations.

The key components are C_8 and R_{16} in the schematic of Fig. 5.10. At power-up, the intermediate node between these two components is tight to ground because of the discharged capacitor. The NAND device U_{6A} connected as an inverter causes the reset signal to be *high*, but when the capacitor charges following V_{cc} grow-up the reset signal is turned *low* thus enabling the whole system. The important point is to choose the RC time constant in such a way that the inverter threshold is reached when V_{cc} is over 4.0 V ensuring a safe operation voltage. You can see this behavior looking at simulation in Fig. 5.13.

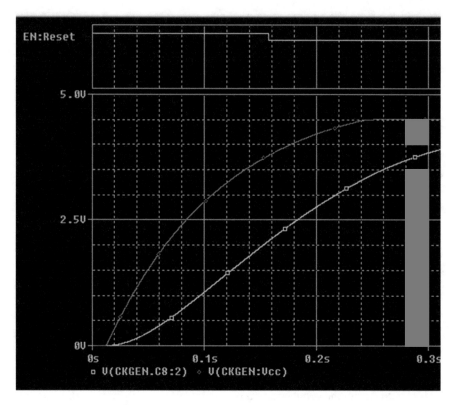

Fig. 5.13 Relevant waveforms of power-on reset circuit

In this figure, you can see V_{cc} in red rising and stabilizing at 4.5 V, the intermediate node (green) tracking V_{cc}, and gradually charging and turning to zero reset signal (in green in the digital section of the figure).

Question for you: *How do you calculate the time constant driving the behavior of the intermediate node? What are the factors that affect this time constant?*

Finally, you can give a look at Fig. 5.14 to have a summary of the internal waveforms we have previously analyzed in the block CKGEN.

You can notice in the upper side of the figure the logic signals that the timing reference of our clock: starting from the bottom you can see the reference pulses obtained starting from the power line with 50 Hz frequency (in violet in the analogic part of the figure). Going up in the digital part, you can see the result of the first division-by-five (CKA) and the following division-by-ten (CKin) which results in a pulse per second that can be sent to the dividers of seconds, minutes and hours. Please note that the division of the primary signal REF starts only when the signal reset has gone low and we are then sure that all the logic of the clock works correctly. In the analogic section of the figure, we can also see in yellow the supply line rectified by the diode bridge with the residual ripple left by the filter capacitor C_9. In red, you

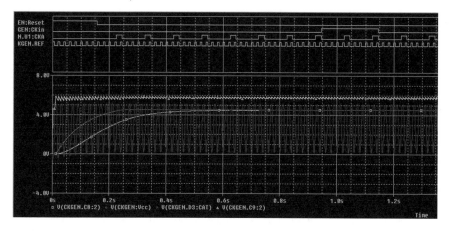

Fig. 5.14 Summary of the internal waveforms of the block CKGEN

then have the full DC supply after the regulator, and finally in green the charge up of the power-on reset node, as we have previously seen.

Let's now turn our attention to the block *COUNT*. Its content is shown in Fig. 5.15.

This block counts up to 59 s and it is divided into two parts: the first one must count the *units* of seconds and goes from 0 to 9, and it is built with the well-known 7490A used as a decade counter with BCD count sequence. The four outputs of this IC are sent to a 7442A which is a decoder IC that makes a correspondence between

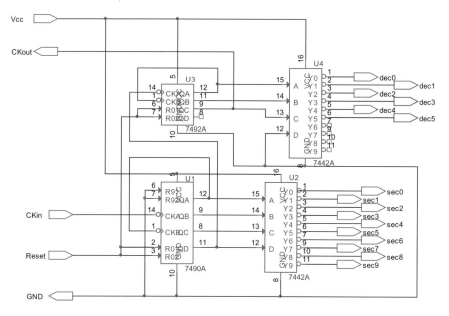

Fig. 5.15 Circuit of block COUNT

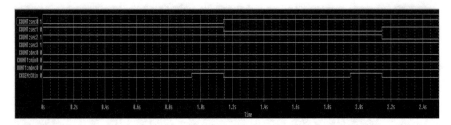

Fig. 5.16   Simulation of the COUNT block for a 2.5 s time interval

BCD configuration and a decimal number represented by an output of the IC. For example, 0010 will correspond to 2 (Y_1), 0111 to 8 (Y_8) and so on. Then, you could be able to display the time directly with decimal numbers (As usual you can get the datasheet of this device to check operation conditions and connections. For all the ICs used in this design, I used the models embedded in OrCAD SW).

This decoding method was used in the past but it is now obsolete because the modern display devices use a combination of seven segments, so the translation of BCD code should result in a combination of a number of signals, nevertheless I choose to maintain it because it is easy to recognize the decimal number looking at the simulation result.

After ten seconds, a pulse is generated and sent to the second part of the block COUNT which counts 0–5 or in other words it must count six *tenths* of seconds. Using the same concept of the decade counter, I have taken a 7492A which is a divide-by-twelve counter that can be also used as a divide-by-six unit which is actually what we need (also for this IC you can find the datasheet).

Then, the outputs of block COUNT will be: dec(0:5) to indicate the tenths of seconds and sec(0:9) to indicate the units of seconds. Please have a look to Fig. 5.16 showing the simulation of block COUNT considering an interval of about 2.5 s because we have to limit simulation duration. For this reason, the ramp-up of the supply voltage has been avoided using temporary voltage generator.

Looking at Fig. 5.16 you can see at the bottom the clock pulse at 1 Hz (1 pulse per second)while starting from the top the outputs sec0-3 are shown. The signals dec0 (0 tenths of seconds), min0 (0 min) and mdec0 (0 tenths of minutes) are also shown. Please verify that at the beginning the signals are set with sec0 low, sec1-3 high and dec0, min0 and mdec0 all low. This means that our clock is set on 00:00:00. After the first second (red line in the figure), sec0 goes high while sec1 goes low as indicated by the logic values on vertical axis. The clock output is now 00:00:01. At the end of 10 s, we would see the signal dec0 going high and the signal dec1 (not shown here) going low and so on.

In this section, we have seen a very common application of counters, but we have also understood how it is possible to manage complex circuits using the hierarchical blocks:

You could try to complete the design of electronic clock adding the module for the hours. Find a way to reset the clock after the number 23:59:59. Using the 50 Hz

clock, it will be far too long to simulate the behavior of the whole clock, then you
should replace the clock pulse coming from CKGEN with a much faster one: do you
think this is possible or not?

5.4 An Adder Circuit

In the binary number system, it is possible to sum two numbers exactly as we do in
the decimal system. Let's take two binary numbers in all the possible combinations
and see the result of a sum:

$$0 + 0 = 0$$
$$0 + 1 = 1$$
$$1 + 0 = 1$$
$$1 + 1 = 10 \text{ (the 1 preceding the 0 in this sum is the carry).}$$

This is exactly what happens with the decimal system, but there the carry comes
when we sum $9 + 1$.

We can translate these relations into a table that can be used to write logic equations
and to syntethize a logic network: the two numbers considered (A, B) will be the inputs
and the sum (S) will be the output. In a similar manner, we can have an equation for
the carry (C). We come out with the following equations:

$$S = AB^* + A^*B \tag{5.5}$$

(the symbol * after the letter indicates the complement of the logic variable)

$$C = AB \tag{5.6}$$

The first equation is a well-known basic logic gate, do you recognize it?
Equations (5.5) and (5.6) define what we call an *half adder*. It is called like that
because it doesn't take care of the carry from a previous stage, which instead we
have to consider in case of addition of multibits numbers. In this case, we should
write a table like this (Table 5.5):

Once again from this table, we can write the logic equation of the new adder that
is called *full adder*. They are:

$$S = ABC_{n-1} + AB^*C_{n-1}^* + A^*BC_{n-1}^* \tag{5.7}$$

$$C_n = BC_{n-1} + AC_{n-1} + AB \tag{5.8}$$

To find these equations, I have used Karnaugh maps.
In particular Eq. (5.7) may be written in another way

Table 5.5 Truth table of a full adder circuit

Carry-in (C_{n-1})	Addend (A)	Augend (B)	Sum (S)	Carry out (C_n)
0	0	0	0	0
0	0	1	1	0
0	1	0	1	0
0	1	1	0	1
1	0	0	1	0
1	0	1	0	1
1	1	0	0	1
1	1	1	1	1

$$S = AC_n^* + BC_n^* + C_{n-1}C_n^* + ABC_{n-1} \tag{5.9}$$

A small exercise for you: *try to obtain* Eq. (5.9) *using (5.7) and (5.8)*.

Well, now it's time to go back in our virtual laboratory and 'build' the circuits described by (5.8) and (5.9).

You can see the drawing in Fig. 5.17.

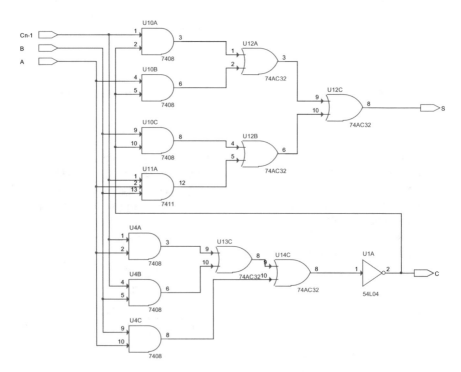

Fig. 5.17 Circuit of a full adder

You can easily verify that this circuit works correctly, applying two binary digits at the input and looking the result on S and C output, but now I would like to use this circuit to build something slightly more complex: a serial adder.

5.5 A Serial Adder

In a serial adder, all the digits of a number are added one after the other using a single adder circuit. The augend and the addend are stored into shift registers and also the sum is loaded in a shift register. You know that shift registers are made of cascaded flip-flops (FFs). Each FF output is connected to the input of the next FF and a common clock pulse is synchronously applied to all the FFs. Basically, a shift register has an external control other than the clock: it is the input of the first FF of the chain and there is an output from the last stage. Then, a binary value can be loaded at the input of the shift register and it will move through the FF chain at each clock pulse until it comes out from the output of the last FF. So, let's see in Fig. 5.18 the schematic of our serial adder

In this schematic, three shift registers are present. They are the IC type 7495A. Also, there is a D-type FF and a sub-circuit named 'adder' which contains the circuit of Fig. 5.17.

You can take the datasheet of 7495A and understand that this IC has a first operation mode in which the data present at the inputs A–D are loaded on the register at the low transition of CLK2 when the pin MODE is set to low, and a second operation mode in which the content of the register is shifted right at each low transition of CLK1 while at the same time, the data at the input SER enters the first FF of the register and moves toward the last one (I invite you to read carefully the 7495A datasheet to further clarify the operation mode of this IC).

This serial adder makes the addition of two numbers of four-bit length. The four bits of each number are loaded in parallel into U_1 and U_2, respectively, then at each CLK1 low transition, two digits are added in the adder block and the resulting bit is loaded in U_3 through the serial input. So, at every clock transition, a new couple of bits are added and the results move through U_3 resulting in a four bits number present on the outputs Q_A–Q_D after four clock cycles which is the result of the sum. Let's know focus on the carry operation: after each sum the generated carry has to be used for the next sum and then must be re-loaded to the input of adder block, but only when the new pair of bits to be added is present. The FF U_{4A} has just the function of 'synchronize' the carry with the next pair of bits.

Well, I have talked a lot, it's time to capture the schematic and prepare the simulation run. We can use the stimulus editor to create the clock pulse, preset the control signals in the initial condition and drive them in the right way during operation. Remember that we have to use initially the shift registers to load the addend and augend so we will have to use MODE and CLK2. We can also load a 0000 on U_3, and we need to preset to zero the output of U_{4A}.

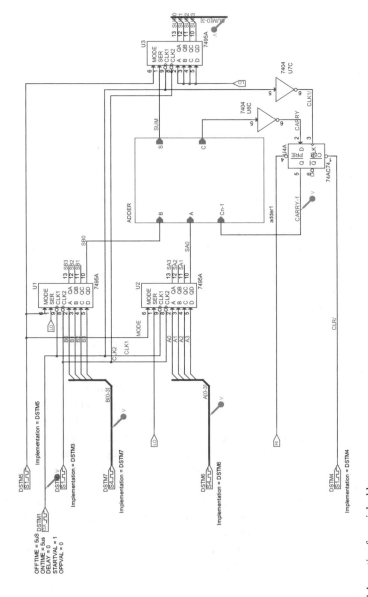

Fig. 5.18 Schematic of a serial adder

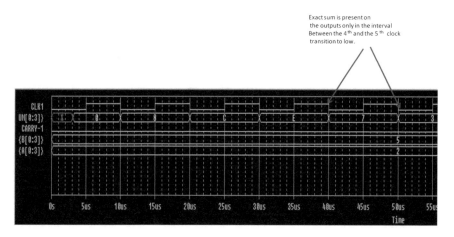

Fig. 5.19 Output of simulation of serial adder

Please note that we have the possibility to give a stimulus on a string of bits (four in our case). This is very useful because we can simply define DSTM6-7 writing a hexadecimal number and use it to drive a 4-bit bus.

I used a bus also to collect the 4-bit data out. It is very easy way to visualize the result as an hexadecimal number. Please have a look to the output plot of the simulation (Fig. 5.19).

In this figure, I have reported the pair of number to be added (A, B) and the result SUM. It is also reported CARRY-1 which is the carry value of operation. You can see here that the use of bus representation makes it possible a very easy reading of the results.

In this example, the operation $5 + 2$ (Hex) is considered. The result is obviously 7. You see from Fig. 5.19 that the result of the operation is available on the register 'SUM' but it is meaningful only after four clock cycles. Actually at that time it appears on the register the correct result: 7. This is because the results of the addition bit by bit of the two input numbers are shifted along the output shift register until all the four digits have been added.

The serial addition is obviously slower than the parallel operation, but it allows to save a number of components.

Before ending this chapter, you can try to sum different numbers. Try to modify the input stimuli for A and B changing them every four clock cycles and verify the correct result on 'SUM' register.

Remember that the result is meaningful only after four clock cycles from the input variations.

Chapter 6
Oscillators

6.1 Introduction

In this chapter, we will perform some experiments with oscillators. They are a very important class of circuits that are used in many key circuit blocks of most of the electronic systems. They are the basic components of radio transmitters for example, but they are the heart of computers also, giving the clock to all the machine operations.

In the beginning, some classical oscillator circuits able to generate a sinusoidal waveform will be considered. Then, logic application of oscillators generating a square wave will be considered, and a clock generator will be analyzed.

6.2 Phase Shift Oscillator

We have seen, speaking about linear amplifiers, that *negative* feedback is a key technique widely used in linear amplifier design. The concept of negative feedback is to subtract a portion of the output signal from the input signal and send to the amplifier the difference of them. This mechanism tends to produce an output waveform perfectly equal to the input, which means to drastically reduce the distortion of the amplifier. At the same time, the stability of the amplifier is greatly increased leading to a larger bandwidth. But what happens if we build a circuit with a *positive* feedback instead? The answer is that such a circuit will be highly unstable because a small discrepancy between output and input signals instead to be reduced will be amplified at each passage in the feedback loop until an oscillation takes place. This behavior should be accurately avoided in a linear amplifier, but it is intentionally searched in the circuits called *oscillators* that are able to generate a sinusoidal waveform of a predetermined frequency.

In Fig. 6.1, you can see a block diagram of a feedback amplifier: The block A is an inverting amplifier with gain $= -A$ so that the relationship between the output V_o and the input V_i is:

© Springer Nature Switzerland AG 2020
R. Gastaldi and G. Campardo, *Electronic Experiences in a Virtual Lab*,
https://doi.org/10.1007/978-3-030-45179-0_6

Fig. 6.1 **Fig. 6.1** Block diagram of a
feedback amplifier

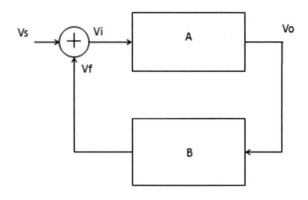

$$V_o = -AV_i \tag{6.1}$$

On the other hand, the feedback signal V_f (in this case a voltage) is related to the output through the gain B of the feedback network so that:

$$V_f = BV_o \tag{6.2}$$

Due to the summing node at the input and the fact that V_o is actually negative, the signal V_i is given by:

$$V_i = \frac{V_s}{(1 + AB)} \tag{6.3}$$

which leads to the classical relationship for the closed-loop gain of a feedback amplifier:

$$\frac{V_o}{V_s} = -\frac{A}{(1 + AB)} \tag{6.4}$$

While the product AB is called *loop gain*.

Now, if we want to obtain an oscillator, first of all we need to design the feedback network B in such a way to introduce a phase shift of $180°$ of V_o, compensating the phase shift introduced by the amplifier A, so that the signal V_f has zero phase shift and is added and not subtracted to the signal V_s. In this way, V_i will be amplified at each loop into the network leading to oscillation...but... you will notice that simply there is no signal input V_s in an oscillator circuit (the block diagram of such a circuit should be better described from Fig. 6.2) so something here seems to be wrong! Don't worry, it is all right because actually do exists an input signal even in an oscillator circuit and it is simply the noise that is present in all the electronic circuits and allows the oscillation to start.

Let's focus a bit on the feedback network: It must be able to induce a negative phase shift; then, it must be done not only with resistances but also with capacitors. Also,

Fig. 6.2 Block diagram of an oscillator

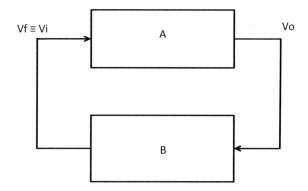

this means that the right phase shift will be obtained at a well-defined frequency of the signal inside the loop, and this is the frequency at which our oscillator will work. Then, changing the resistive and capacitive elements inside the feedback network, we are able to design an oscillator working at the frequency we want (obviously in the limits of network characteristics).

Now I think you will rise another point: If the signal is continuously amplified in the loop, what we will obtain is something that will exceed the amplifier dynamic range ending to crash against the power supply limit and not a stable sinusoidal waveform. I totally agree with this remark, the key point here is the loop gain AB: If AB > 1, the result is an exponential increase leading to the result said above, but AB < 1 is not a solution because in this case the oscillation will set but it will be weaker and weaker at each loop until it stops. To guarantee a stable oscillation, we need to ensure the condition AB = 1

Our conclusion then is that to design an oscillator, we need a feedback circuit in which the gain loop meets the following conditions:

$$AB_{\omega osc} = 1 \tag{6.5}$$

$$\varphi(AB)_{\omega osc} = 0 + 2K\pi \tag{6.6}$$

These two conditions are called *Barkhausen conditions*, and they are met at a certain frequency of the signal which is the oscillation frequency.

Now let's go to our virtual laboratory and enter the schematic of Fig. 6.3.

In these schematics, the amplifier stage A is the bipolar transistor (BJT) Q_1 which is connected in common emitter configuration. Biasing the base of BJT is obtained through the resistors R_4 and R_5, while an emitter resistance R_7 introduces a series–series feedback in the circuit which stabilizes the operating point. We know that the voltage gain of amplifier A is approximately given by the ratio between resistances R_3 and R_7.

The feedback network B is built with C_1, R_1, C_2, R_2 and capacitor C_3 which sees the input resistance (R_{in}) of the amplifier stage. Theory tells that:

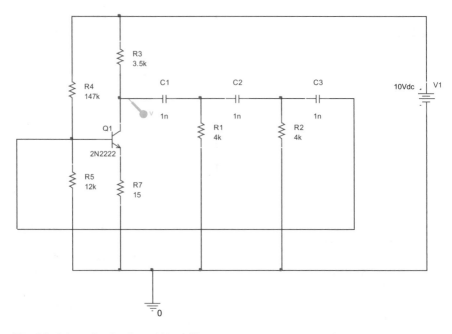

Fig. 6.3 Schematic of a phase shift oscillator

$$R_{in} = h_{ie} + (1 + h_{fe})R_7 \tag{6.7}$$

where h_{ie} is the input resistance of the BJT in the linearized small-signal model and h_{fe} is the BJT current gain:

$$h_{ie} = \frac{h_{fe} V_T}{|I_c|} \tag{6.8}$$

In the above definition, I_c is the bias collector current of Q_1 and V_T is the so-called volt equivalent of temperature. At ambient temperature, $V_T = 26$ mV. We can choose the values of components and the BJT bias point to have R_{in} ~4 K. Can you recognize the block schematic of Fig. 6.2? I think yes! Now we can use the Barkhausen conditions to design the circuit to produce a stable oscillation and to predict which is the *frequency* of such an oscillation. We can break the feedback loop and evaluate the loop gain. Then, C_3 must be disconnected from the base of Q_1, but we have to add the impedance (R_{in}) it saw when it was connected so that the network remains unaffected. We can connect a signal generator V_s to the base of Q_1 to calculate open-loop gain (see Fig. 6.4).

In the AC feedback network, both the sampled signals are voltage and the feedback signal is a current, and then, the open-loop gain AB($j\omega$) can be written as:

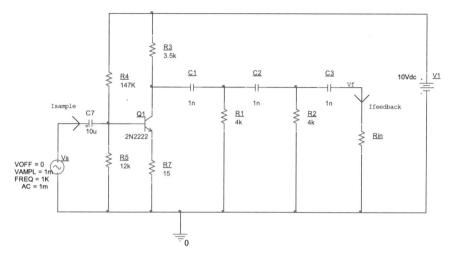

Fig. 6.4 Circuit of the oscillator with the feedback loop broken at the input of the amplifier. Capacitor C_7 has been added to avoid DC bias point to be disturbed by the injected signal

$$AB(j\omega) = \frac{I_{\text{feedback}}}{I_{\text{sample}}} = \frac{V_f/R_i}{V_s/R_i} = \frac{V_f}{V_s} \tag{6.9}$$

The theory gives the expressions for the module $|AB(j\omega)|$ and the phase $\varphi_{AB}(j\omega)$ in the case in which $R_{\text{in}} = R_1 = R_2 = R$ and $C_1 = C_2 = C_3 = C$(Ref. Fig. 6.4).
It comes out that the frequency of oscillation is given by:

$$f_0 = \frac{1}{2\pi\,RC\,\sqrt{6+4k}} \tag{6.10}$$

where

$$k = \frac{R_c}{R} = \frac{R_3}{R} \tag{6.11}$$

Let's substitute to this general relationship the real values of the schematic in Fig. 6.4 we can compare the number we have now with the result we will have from simulation.
We obtain then

$$f_0 = \frac{1}{2\pi\left(4\times10^3\right)\left(1.\times10^{-9}\right)\sqrt{6+3.5}}$$

The result is $f_0 = 12.9$ kHz.

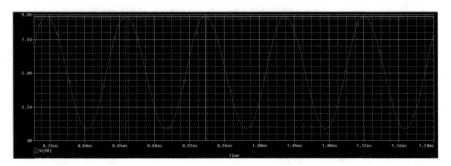

Fig. 6.5 Frequency measurement of the sinusoidal waveform coming from the circuit of Fig. 6.3

Now, all this stuff was a little boring, but we hope it contributed to clarify the mechanism of operation of an oscillator. Then it's time to simulate our circuit and compare the results with theoretical predictions.

Coming back to the circuit of Fig. 6.3, the result of a simulation run is shown in Fig. 6.5 where we can see the oscillator running in a sinusoidal waveform. If we measure the distance in time between two peaks, we find a frequency of about 11 kHz that is about 15% less than what we found with the theoretical formula.

In Fig. 6.5, now it is interesting to investigate the impact of the gain of the amplifier stage using Q_1 on the oscillator waveform: According to the theory, we expect that by reducing the gain of the amplifier, the stable oscillation will turn to a damped one.

One easy way to reduce gain is to increase the value of emitter resistance. Let's increase the R_5 up to 100 Ω. The corresponding result of simulation in this condition is shown in Fig. 6.6.

You can easily see a sinusoidal waveform whose amplitude is rapidly decreasing with time ultimately reducing to zero, and then, we can conclude that a stable oscillation cannot take place. This is in accordance with the prediction of theory as in this case open-loop gain is lower than 1 at the oscillation frequency.

To complete our analysis, let's try now with R_{17} decreased to 10 Ω.

The result is shown in Fig. 6.7: As a result of increased gain, the oscillation amplitude has increased but at the price of a high distortion, due to nonlinearity of the amplifier at that amplitude.

You can now take the circuit of Fig. 6.4 in which the AC feedback loop is broken to see the behavior of open-loop gain with frequency. Let's comment the AC sweep of Fig. 6.8 obtained with this circuit: In the upper part of the figure, you can see the module of loop gain, in the lower part you can see the phase. Now let's find the Barkhausen condition: Open-loop gain must be at least one when the phase shift is zero. The point is marked with a vertical red line in Fig. 6.8, and we can read the corresponding frequency which is the oscillation frequency: It is 11 kHz in agreement with our previous simulation.

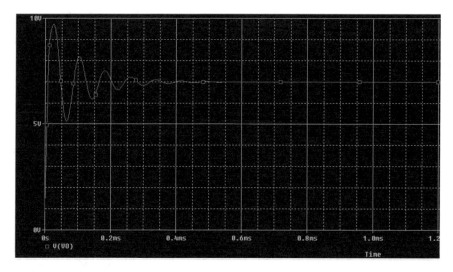

Fig. 6.6 Oscillator of Fig. 6.3 running with R_{17} increased up to 100 Ω

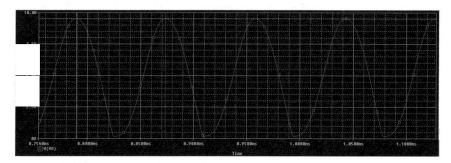

Fig. 6.7 Oscillator of Fig. 6.3 running with R_{17} reduced to 10 Ω

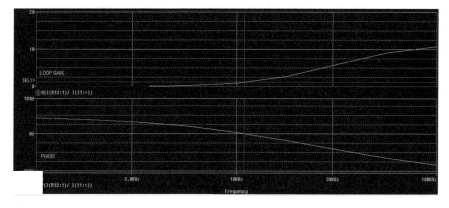

Fig. 6.8 Bode plot of the open-loop gain for the phase shift oscillator (Ref. Fig. 6.4)

Before turning to the next section, please elaborate a bit on the following item:
The phase shift network must provide a 180° shift. Why we need three RC cells to achieve this result?

6.3   An Example of Tuned Oscillator

In the previous experiment, the key to generate oscillations was a feedback network capable to apply to the input of an amplifier stage a signal with an appropriate phase shift respect to the output, instead in the next experiment the feedback network is a resonant circuit. The effect of this network is that only a signal at the resonance frequency can propagate in the loop resulting in an oscillator at that frequency.

A resonant circuit is made of capacitor and inductor connected in parallel so that if we provide energy to the system, for example charging the capacitor, then energy will be transferred from capacitor to inductor transforming from electrostatic field energy to magnetic field energy and vice versa. In an ideal system, this back and forth behavior will continue endless (for this reason, this network is also called *tank* to remember that it can in some way store the energy provided to it) and the circuit will oscillate at a well-defined frequency called *resonance frequency* but of course in the real world, joule effect in wires and electromagnetic radiation emission will slowly dissipate the initial energy, so that what we can observe is a dumped oscillation.

Anyway, if we use a resonant circuit as a feedback network of an amplifier, we can obtain a stable oscillation exactly at the resonance frequency. Two very famous circuits using this principle are the *Colpitts* and the *Hartley* oscillators. The main difference between the two is that the role of capacitors and inductors in the tank is exchanged. Now we are going to work with a Colpitts oscillator, while in a next chapter, you will see a Hartley circuit as a part of a more complex system.

Now please enter on your simulation tool the circuit of Fig. 6.9.

In Fig. 6.9, the tuned network is made of L_1, C_4, C_5. The output is taken between C_1 and C_4. The oscillation frequency is given by the relationship:

$$f_0 = \frac{1}{2\pi \sqrt{L_1 C_{\text{eff}}}} \tag{6.12}$$

where

$$C_{\text{eff}} = \frac{C_4 C_5}{C_4 + C_5} \tag{6.13}$$

Let's substitute the real numbers in these equations:

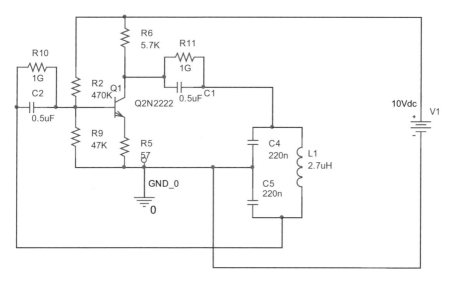

Fig. 6.9 Colpitts oscillator schematic. Please note that R_{10} and R_{11} are not really present in the circuit. They are added to avoid that the simulator consider L_1 floating

$$f_0 = \frac{1}{2\pi \sqrt{110 \times 10^{-9} \times 2.7 \times 10^{-6}}}$$

$$C_{\text{eff}} = \frac{220 \times 10^{-9} \times 220 \times 10^{-9}}{220 \times 10^{-9} + 220 \times 10^{-9}}$$

$C_{\text{eff}} = 110 \text{ nF}$
$f_0 = 292.226 \text{ kHz}.$

Now just launch the simulation looking at the signal in the point indicated by the probe in Fig. 6.9.

The result is shown in Fig. 6.10, and it is easy to verify that our prediction of oscillation frequency was correct. The period shown in Fig. 6.10 is 3.4513 uS which corresponds to a frequency of about 290 kHz.

The frequency of oscillation can be changed varying the values of $C_4 - C_5$ or L_1.

Before closing this section, we would like to propose a question for you:

Is it better to use C or L to change the oscillation frequency? Can you specify why?

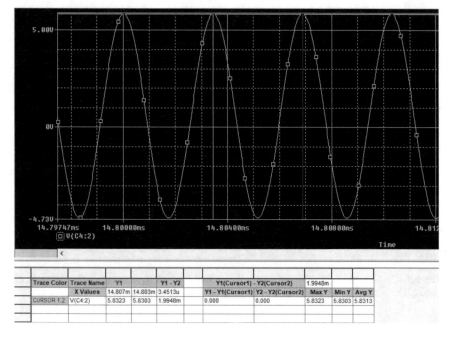

Trace Color	Trace Name	Y1		Y1 - Y2		Y1(Cursor1) - Y2(Cursor2)		1.9948m			
	X Values	14.807m	14.803m	3.4513u		Y1 - Y1(Cursor1)	Y2 - Y2(Cursor2)	Max Y	Min Y	Avg Y	
CURSOR 1,2	V(C4:2)	5.8323	5.8303	1.9948m		0.000	0.000	5.8323	5.8303	5.8313	

Fig. 6.10 Output of Colpitts oscillator. It is visible in the period = 3.4513 uS

6.4   Schmitt Trigger as a Clock Generator

The Schmitt trigger is a very famous logic circuit that you can find in a variety
of electronic systems. You certainly know that it is a comparator with a *hysteresis*
which means that the threshold going *up* is different from the threshold going *low*.
The hysteresis curve can be schematized in Fig. 6.11.

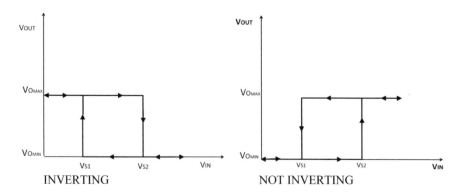

Fig. 6.11 Hysteresis characteristics of a Schmitt trigger

In Fig. 6.11, the hysteresis is defined as $V_{S2} - V_{S1}$ and the operation can be not-inverting type, which means that the output of the circuit goes high level when input crosses the higher threshold V_{S2} and switches down to low level when input comes back through the lower threshold V_{S1} or it can be the opposite, called inverting type.

This characteristic is a key factor to clean the noise associated with the input waveform and makes it possible to fabricate reliable control circuits. For example, let's suppose to design a ON-OFF temperature controller: As soon as the temperature reaches the reference value, a switch goes off interrupting the power to the heater, but if we operate with a single threshold it will be enough a very small drop in the temperature maybe caused by some noise to set the switch ON again, but the power delivered to the system will be enough to overcome the reference temperature determining a new switch OFF. In the end, the system will continuously oscillate between one state and the other and will not operate correctly. Instead if two thresholds are available, the temperature needs to drop below the second (low) threshold before the heater will be powered again. In this way, the system can work in a stable mode. On another side, hysteresis can be used to generate a square wave starting from a variable signal as an input to Schmitt trigger: The only condition is that the input signal crosses the upper and lower threshold voltages of the circuit. The circuit I want to use is shown in Fig. 6.12, and it is of not-inverting type. Before using it to obtain a squared wave (actually the result we obtain is a waveform with a duty cycle generally different from 50%, depending on the input signal), we will find the theoretical high and low threshold levels and will verify them with simulation.

The theoretical value of the thresholds for ON and OFF switching points is given by the following relationships:

$$V_{s1} = V_{en} + V_{\gamma} \qquad (6.14)$$

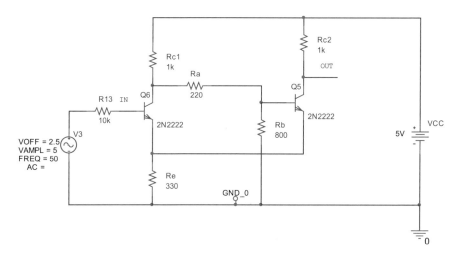

Fig. 6.12 Use of a Schmitt trigger to generate a square wave starting from an arbitrary waveform. In the figure, it is indicated also a sine wave generator used to simulate the behavior of the circuit

$$V_{s2} = V_{be} + R_e \times \frac{1}{(R_e + aR)}(V' - V_\gamma) \tag{6.15}$$

Let's give some definitions:

$$V_{en} = (V' - V_{be}) \times R_e \times \frac{(h_{fe} + 1)}{(R_b + R_e(h_{fe} + 1))} \tag{6.16}$$

$$V_\gamma = 0.5 \, \text{V(cut-in } V_{be}\text{)}$$

$V_{be} = 0.6 \, \text{V}$ (base-emitter voltage to have a transistor fully ON).

$$R_b = \frac{R_2(R_{c1} + R_a)}{(R_{c1} + R_a + R_b)} \tag{6.17}$$

$$V' = \frac{V_{cc} \times R_b}{(R_{c1} + R_a + R_b)} \tag{6.18}$$

$$R = \frac{R_{c1} \times (R_a + R_b)}{R_{c1} + R_a + R_b} \tag{6.19}$$

$$a = \frac{R_b}{R_a + R_b} \tag{6.20}$$

I realize all this stuff is quite boring, but I would like to compare the results given by theory for ON and OFF points which are reasonably in accordance with the results given by simulation. Nowadays, CAD tools are very powerful and precise, but it is of fundamental importance to be able to make approximate evaluations of the results utilizing when possible 'simple' formulas to be sure that simulation is going on the right way and we didn't make mistakes giving the inputs to the SW.

Then, let's substitute the values of components of the circuit in Fig. 6.12 in the definitions above, remembering that we consider $V_{be} = 0.6$ V for silicon and $h_{fe} = 100$ and we obtain $V_{S1} = 1.82$ V and $V_{S2} = 1.28$ V. Let's also keep in mind the value of $V' = 1.98$ V.

The result of the simulation of the circuit shown in Fig. 6.12 is reported in Fig. 6.13, where it is clear that the output goes to high level as soon as the sine wave in input which spans from 0 to 5 V crosses the first threshold V_{S1} and goes back to low level when the input is lower than V_{S2}. The result is a square wave with the same period as the input sine wave but with a duty cycle different from 50%.

The two thresholds are set at $V_{S1} = 1.82$ which is exactly the value predicted with the theory, while $V_{S2} = 1.40$ which is slightly higher than expected using our formulas (see Fig. 6.14). The important point with Schmitt trigger is that if h_{fe} is big enough, the threshold of the circuit is determined only by the resistors of the circuit and it is practically independent from the characteristics of the transistors.

Using a Schmitt trigger circuit, it is possible to build a square wave generator (practically an oscillator) controlling also the duty cycle. It is done simply using

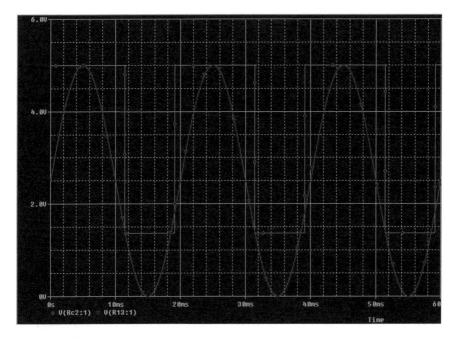

Fig. 6.13   Schmitt trigger of Fig. 6.12 used to create a square wave starting from a sine wave as an input

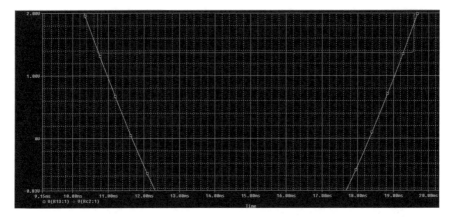

Fig. 6.14   Detail of the trigger levels of the Schmitt trigger

a capacitor charged by a resistance connected to the output of the Schmitt trigger. An example of such a circuit is shown in Fig. 6.15, where you can actually see a three-stage circuit comprising a Schmitt trigger, an inverter and an emitter follower as the last stage.

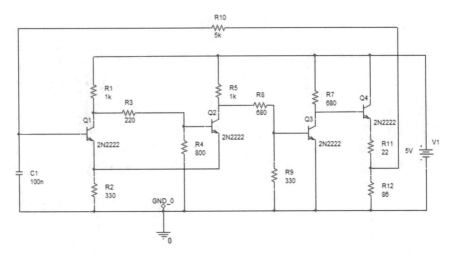

Fig. 6.15 An oscillator generating square waves made with a Schmitt trigger

The concept of operation is simple: The capacitor starts to charge until it reaches the upper threshold voltage, and then, the Schmitt trigger switches; now the capacitor starts *to discharge* until the lower threshold is reached, then the cycle starts again. The result of simulation shown in Fig. 6.16 confirms what we have said. From Fig. 6.16, you can clearly distinguish the sawtooth waveform across the capacitor, the square

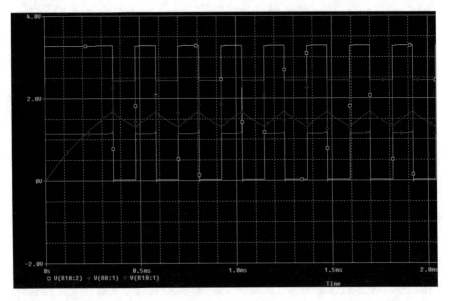

Fig. 6.16 Simulation of the square wave generator. You can see the voltage across the capacitor (blue), the output of the Schmitt trigger (red) and the output of the oscillator (green)

Fig. 6.17 Equivalent circuit
of a crystal resonator

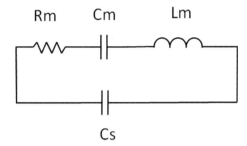

wave at the output of Schmitt trigger and the same with full logic levels at the output
of the oscillator.

Let's finish this section with the usual 'quiz.' Please try to answer the following
questions; it should be easy after what we have said.

- *What are the components that determine the period of the square wave? And duty
  cycle?*
- *Why it is necessary to insert an inverter stage?*
- *What is the function of the final emitter follower? What happens if we remove it?*
- *It is possible to modify the feedback network to have an adjustable duty cycle?*

6.5   Crystal Oscillator

When it is needed an oscillation very stable in frequency a common solution is to use
a quartz crystal in the feedback loop. The crystal behaves as a tuned network which
can be schematized as in Fig. 6.17. The values of components change a bit from a
piece to the other. Every crystal oscillates at a well-defined frequency determined at
the moment of fabrication. The big advantage of using quartz in a circuit is that the
oscillation frequency is quite independent from the variations of parameters of the
circuit and of the environment such as the temperature.

You can see the circuit in Fig. 6.17 is a bit complex; the components shown have
the following significance:

R_m represents the mechanical losses in the crystal. These have to be compensated
by the amplifier and external network to maintain a stable oscillation.

C_m represents the electrical charge gained during vibrations in the crystal

L_m takes into account the movement of the crystal mass.

Finally, C_0 represents the shunt capacitance between the electrodes of the crystal
plus the stray capacitance of the holder.

The crystal has two characteristic frequencies: The first is the series resonance
frequency at which L_m and C_m compensate each other and the total impedance is the
parallel of R_m and C_s, but the reactance of C_s is very high at the frequency considered,
so the equivalent impedance is approximately a pure resistance R_m.

The series resonant frequency F_s can be obtained from the following relationship:

$$F_s = \frac{1}{2\pi}\sqrt{\frac{1}{L_m C_m}}$$ (6.21)

The other characteristic frequency is said anti-resonant frequency and is given by the following equation:

$$F_A = \frac{1}{2\pi}\sqrt{\left(\frac{1}{L_m C_m} + \frac{1}{L_m C_s}\right)}$$ (6.22)

In the frequency region between F_S and F_A, the crystal operate as an inductive reactance, while outside this region it appears as a capacitive reactance. This region is also known as parallel resonance region. Crystal operation is placed between F_S and F_A which means in the region of inductive reactance.

Let's now have a look to the circuit of Fig. 6.18.

The configuration shown in this figure is called Pierce oscillator and is simple to understand: M_1 and M_2 are MOS transistors connected to form an inverter, the

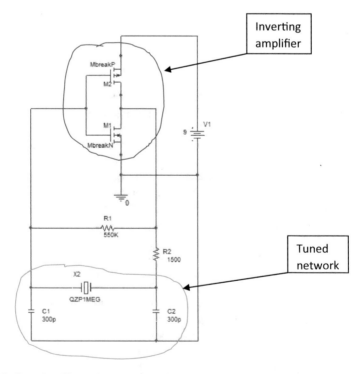

Fig. 6.18 Crystal oscillator Pierce configuration

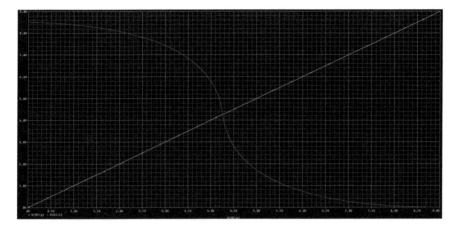

Fig. 6.19 Output characteristic (red curve) of the inverter amplifier. The green line is corresponding to the condition $V_{in} = V_{out}$ crossing the characteristic about half of the supply voltage and determining the bias point of the circuit due to R_1. In that region, the red curve is almost linear and the gain is maximum

driver being the NMOS and the load PMOS. Usually, this configuration is used to process logic signals that can assume the value of supply voltage or ground, but in this case the resistor R_1 forces the circuit in a bias condition $V_{in} = V_{out}$; this condition corresponds to a point around half of the supply voltage in the linear portion of the transfer characteristic where the gain is a maximum (see Fig. 6.19), and then, the inverter behaves as a linear inverting amplifier which means that the output voltage has a 180 degrees phase shift with respect to the input. An additional 180° phase shift is provided by the network $C_1 - C_2$ crystal, so that the oscillation can start.

The actual oscillation frequency depends also from the capacitive load seen and then also from C_1 and C_2 as shown by the formula:

$$F_o = F_S \left(1 + \frac{C_m}{2(C_S + C_L)} \right) \tag{6.23}$$

where

$$C_L = \frac{C_1 C_2}{(C_1 + C_2)} \tag{6.24}$$

The role of R_2 is to limit the output of the amplifier so that the crystal is not overdriven. Actually, overdrive can cause a fast aging of the crystal. Its value should be chosen taking into account the crystal manufacturer data on maximum driving capability of the device.

We can now run the simulation obtaining at the output of amplifier-inverter the signal shown in Fig. 6.20.

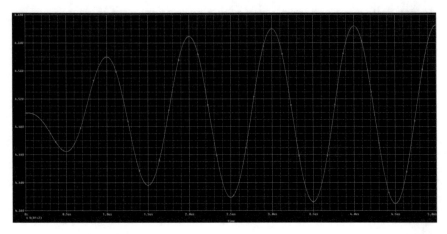

Fig. 6.20 Output of Pierce oscillation using a 1 MHz quartz crystal. You can verify the stable period of oscillation exactly at 1 MHz after the startup

In Fig. 6.21, you can verify that the phase between input and output signals to the inverter is actually shifted to 180°.

As usual let's finish this section with some questions you can try to answer for fun.

– *In the circuit of* Fig. 6.18, *what happens if we change the value of* R_1? *And if we change* $C_1 - C_2$?
– *In your opinion, the circuit of* Fig. 6.18 *can be efficiently used as a clock generator in a digital system*?
– *Check the spectrum of the output signal.*

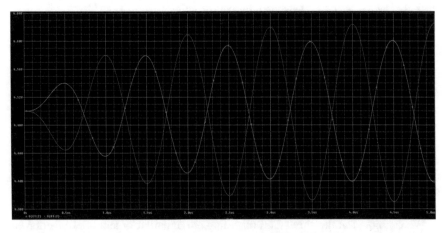

Fig. 6.21 Input (green) and output (red) signals to the inverting amplifier of the Pierce oscillator

Chapter 7
Some Examples

7.1 Introduction

This chapter is a collection of three examples of complex circuits. We start with a photocell application, then a simple circuit to measure the heartbeat and, at the end, a FM transmitter. They are a good example of how a complex electronic circuit is designed and simulated.

7.2 Detection Circuit Driven by a Photocell

The circuit we would like to study and simulate is a simple circuit, used many time for different applications. The reason why we choose this circuit is because it contains a concept often used in the circuit driven by a sensor, like a thermo-actuator, to avoid spurious commutations.

We are speaking about the hysteresis concept, used in Schmitt trigger circuit.

Figure 7.1 shows what normally happens when we need to compare a signal changing in time, the continuous black line, versus a reference, stable in time, the

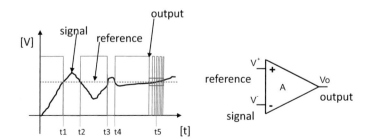

Fig. 7.1 Motivations to use a Schmitt trigger

© Springer Nature Switzerland AG 2020
R. Gastaldi and G. Campardo, *Electronic Experiences in a Virtual Lab*,
https://doi.org/10.1007/978-3-030-45179-0_7

dotted black line. The comparator output will go up and down reflecting the fact that signal is greater or lower than the reference. This is, for example, the principle of the thermo-actuator control. You can set up the temperature, this is the reference, and the system will be on or off depending on the temperature changing. The output is high (in voltage) if the signal is lower than the reference (if we need the opposite it is just enough to reverse the input of the operational amplifier).

When the signal reaches the value of the reference, the output goes low.

And this could be the first problem because we have to take in account the inertia of the system. But this is not the question now.

You can see in the graphics that all works fine for t_1, t_2, t_3 and t_4. What happen around t_5? I said around because we have to think that the operational amplifier we are using is not an ideal OA. Like all the systems in the world, it has a margin of uncertainty. We design, in Fig. 7.1, a rectangular box around t_5. Inside this box, the signal and the reference are very close, much more than in the previous points of the graph. The system (the OA) is no more able to distinguish if the reference is up or down with respect to the signal, so the output oscillates.

This is not a good way to control the system. What we can do to have a better control on the output switch? We need something ables to eliminate the indeterminacy area.

The problem exists if the signal moves with a slope similar to the reference slope in Fig. 7.1. In this situation, signal and reference are close each other, and the uncertainty problem could explode. The solution is to have two reference levels instead of one, as shown in Fig. 7.2.

Figure 7.2 shows the signal, the old reference (dotted line in the middle of the two references) used in Fig. 7.1 and the two new references. The reference's position, as the uncertainty boxes around the reference dotted line, is exaggerated to make it clear to the reader.

And now? It seems that we have not solved the problem but only doubled it. No, we have to specify how to use the two references. We need to design a system able

Fig. 7.2 Two reference solution

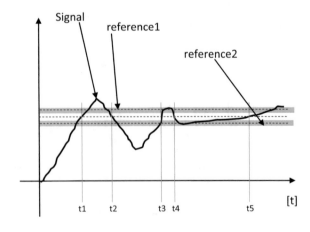

Fig. 7.3 Schmitt trigger function

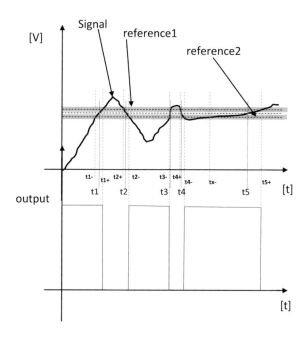

to switch up and down with different thresholds. In Fig. 7.1, we have the output at high value when the reference is higher than signal. Figure 7.3 shows that the output commutes in t_{1+}, going low when the signal crosses the reference 1, greater than the old reference. The output remains low until t_{2-} when it crosses the reference 2 and then stay high till t_{3+} and again goes high at t_{4+} and remain high in the zone where we had the oscillation in Fig. 7.1. The separation between the two reference thresholds is called hysteresis from the Greek word that means 'delay.'

So we have a system able to commute 'uphill' with respect to a reference and 'downhill' with respect to another one. In this way, we obtained a commutation not in the perfect position with respect to the signal and the reference, but, in any case, this is not possible due to the uncertainty, and so, it is better to have a difference in the commutation precision respect to have oscillation. If your thermo-control heaters in your home turn on and off continuously, your boiler will be broken in a short time.

Let's start to see the circuit and then we can continue to discuss, as shown in Fig. 7.4.

The first block includes R_9 that is a LDR, light-dependent resistor, a resistor that changes is value depending of the light exposure; in the presence of light, the value of the resistor decreases and increases with the darkness. The LDR (R_9) resistance varies from 5 kΩ to 500 kΩ. In the second block, R_8 and R_7 are representing a potentiometer used to define the value of the voltage switch for the OA and in this case the level of the light. The Schmitt trigger is composed by the OA, of course, and the resistors R_1, R_4 and R_3. R_2 is used to drive the BJT with the right current. The bipolar transistor Q_1 doesn't work as a linear amplifier but as a switch. R_2 drives the BJT with the right current when it is in saturation (switch closed), the LED diode D_1

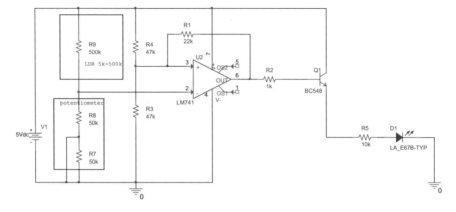

Fig. 7.4 Automatic light control

is ON when the Q_1 is closed. R_5 is needed to limit the diode current, preventing it to burn. To understand how the Schmitt trigger works it is better to imagine to be in a stable condition, for example, with the output high. In this condition, the circuit for the OA is equivalent to Fig. 7.5.

The R_1 will be 'as' connected at VDD so, the V^+, we use like a reference will be at

$$V^+ = \text{VDD}\frac{R_3}{R_3 + R_4/R_1} = \text{VDD}\frac{47\text{ k}\Omega}{47\text{ k}\Omega + \frac{47\text{ k}\Omega * 22\text{ k}\Omega}{47\text{ k}\Omega + 22\text{ k}\Omega}} \sim 3.8\text{ V}$$

This means that when the output is high, the system will recognize the input to turn off the led, with a threshold of 3.8 V.

The input must go below 3.8 V to turn off the diode again.

When the output is low, Fig. 7.6, R_1 will be 'as' connected at GND so, the V^+, we use like a reference will be at

Fig. 7.5 OA condition when the output is high

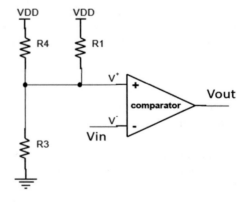

Fig. 7.6 OA condition when
the output is low

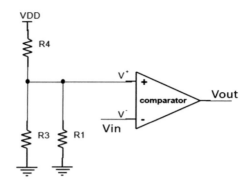

$$V^+ = \text{VDD}\frac{R_3/R_1}{R_4 + R_3/R_1} = \text{VDD}\frac{\frac{47\ k\Omega * 22\ k\Omega}{47\ k\Omega + 22\ k\Omega}}{47\ k\Omega + \frac{47\ k\Omega * 22\ k\Omega}{47\ k\Omega + 22\ k\Omega}} \sim 1.2\ \text{V}$$

This means that when the output is low, the system will recognize the input to turn on the led, using a threshold of 1.2 V.

The input must go below 1.2 V to turn on the diode again.

We said before that the photocell resistance range is from 5 kΩ to 500 kΩ: 5 kΩ with high luminosity, 500 kΩ with the dark. The behavior of the photocell is not 'linear,' is not like a switch but seems more a hyperbole.

Find on the Web a datasheet for a photocell and understand how the output signal moves.

Considering Fig. 7.7, we replace the photocell and the potentiometer, with a PWL (piece-wise linear) generator, to simplify the simulation.

Figure 7.8 shows a green probe and a red one. Green is the input.

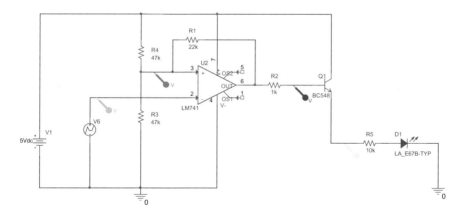

Fig. 7.7 Photocell circuit to simulate

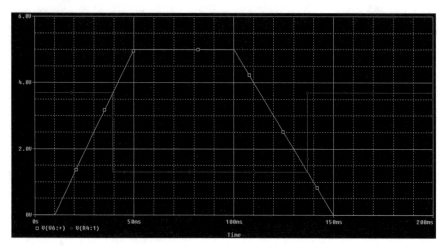

Fig. 7.8 Photocell circuit simulation signals $V_{in} = V^-$ and V^+

It starts from zero that means high value for the photocell resistance, so the dark. Red probe is at ~3.7 V; when Vin reaches the same voltage, we have the switch and the red probe voltage goes low at ~1.3 V. Supply voltage is 5 V, as from Fig. 7.7.

Studying Fig. 7.9, with the blue probe, the output of the OA, we can understand why the threshold levels, we calculated before, are not exactly the same as the simulation results.

I can say why but, probably, it will be better if you find the answer by yourself.

Finally, Fig. 7.10, with all the probes, is shown.

We can do many other simulations to study how the circuit works and try to find its limits.

The first analysis could be done studying how the circuit responds with respect to the photocell switching speed.

Our simulation was done with a millisecond time range. But we can explore what happen if we increase the speed of the luminosity change.

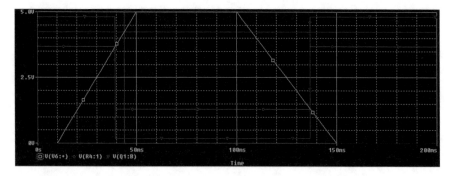

Fig. 7.9 Photocell circuit $V_{in} = V^-$ and V^+ and the OA output, blue probe

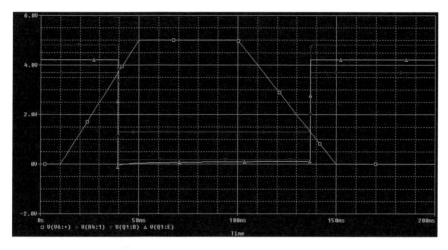

Fig. 7.10 Photocell circuit simulation with all the probes

Try to find this limitation and explain what happens and why.

Another interesting point to be studied is the behavior of the system with respect to a noise, a disturb, on the supply voltage node.

What will happen if, for any reason, the supply voltage will change?

Can we define a maximum value of the supply voltage change that will not disturb our circuit?

How we can increase this value?

7.3 Heartbeat Detector

The second circuit we will address and simulate is a heartbeat detector.

Again, studying this circuit we will explore a concept that is one of the most important in electronic and it is related to an important circuit configuration. We are speaking about the differential stage and the CMRR, the common-mode rejection ratio.

We already saw this circuital configuration, the differential amplifier, that is the core of the operational amplifier, and studied it, but we think that this configuration contains a lot of basic ideas, so that it is right to deeply understand his behavior by studying different applications for this configuration.

An ECG, an electrocardiograph, is used in medical work to examine the electrical pulses produced by the heart. We need to have electrodes, placed on the body, to pick up the signals coming from the heart.

Noises and hums are present and induced by the human body, and the electrical signals we want to get are really little as voltage amplitude.

We are speaking about signal from 10 μV to 3 mV range with DC signals overlying, up to 300 mV.

These DC signals are, for this application, to be considered noise.

We have to extract a very little signal, embedded in a large 'sea' of noise. Like being able to see a little wave, a ripple, embedded in a giant wave.

All the disturb signals, if they can be supposed applied to both the inputs, can be neglected.

For example, signals come from network interference at 50–60 Hz, monitors, ceiling lights or other electromagnetic interferences.

Total voltage present on the electrodes could be of some hundreds of millivolt, to detect the signal we need a differential amplifier using the concept that the noise, all the noises present on the electrodes will be present on all the electrodes and we will detect the differences between them, the heartbeat signal.

What we have to design is a system able to manage low signals and reject hum and noise that, we can imagine, are applied to both the inputs.

A differential amplifier will be used, as shown in Fig. 7.11. Before to simulate the circuit in, Fig. 7.11, the LM741 datasheet shows, Fig. 7.12, for the voltage gain, a typical value of 200 V/mV (200 V/mV = 200,000 V/V). Means that, if we have a differential signal equal to 10 μV, the output will be

$$A_v = 200,000 = \frac{V_{\text{out}}}{V_{\text{in}}} = \frac{V_{\text{out}}}{10 \times 10^{-6}}$$
$$V_{\text{out}} = 200,000 \times 10 \times 10^{-6} = 2 \text{ V}$$

Simulation of circuit in Fig. 7.13 shows a differential gain of 106 dB that means 200,000 like we read for the typical value in the datasheet, with a pole at 4.5 Hz, as shown in Fig. 7.14.

Figure 7.15 shows an open-loop application. The OA is used to have an output equal to the inputs difference amplified. Capacitors C_5 and C_6 increase the input impedance and stop the DC signals.

Now, to complete this very simple heartbeat amplifier, we would like to do something to stop also higher-frequency signals.

So, come back to Fig. 7.11 and see Fig. 7.16 where we put two schematics.

Schematic on the left shows the circuit with the two inputs, same signals with 60 μV of amplitude and a DC generator of 300 mV. The schematic on the right is the copy with only differential signal generator on the input with 120 μV of amplitude.

Simulation done is showed the output signals V_{out} and V_{outx}, Fig. 7.17, violet and blue probes are superimposed and they are the same. This is because the input capacitors stop the 300 mV DC.

We can imagine a disturb in AC like in Fig. 7.18. V_{18} represents an AC noise of 100 mV common to all the input.

In input, we have 1 mV signal.

Please observe carefully the input generators and their polarity.

Again, Fig. 7.19, same configuration without any noise, imagine an ideal differ-

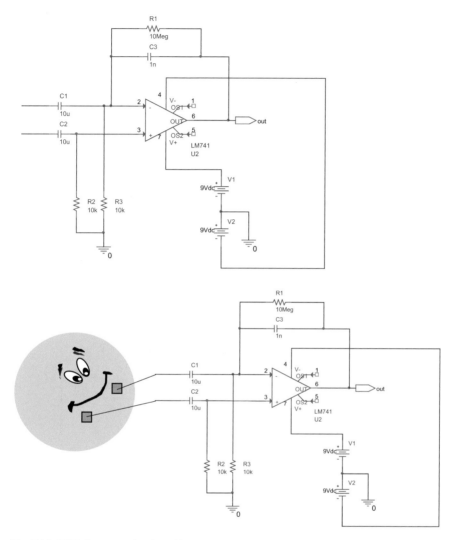

Fig. 7.11 ECG first-stage circuit and how to connect it to our patient

ential configuration. And finally, Fig. 7.20, with both DC and AC common noise generators.

Figure 7.21 compares the three different outputs. We can see that the three signals are very closely one with the other. The CMMR concept works fine.

In Fig. 7.23, we have the Bode analysis for the circuit in Fig. 7.22. Cursors show that at 2 Hz, the gain is ~1250.

For this circuit, many other analyses could be suggested, as:

Simulate the PSRR.

Simulate the variation of the CMRR with the temperature.

6.7   Electrical Characteristics, LM741C[1]

PARAMETER		TEST CONDITIONS		MIN	TYP	MAX	UNIT
Input offset voltage	$R_S \leq 10\ k\Omega$	$T_A = 25°C$			2	6	mV
		$T_{AMIN} \leq T_A \leq T_{AMAX}$				7.5	mV
Input offset voltage adjustment range	$T_A = 25°C$, $V_S = \pm20\ V$				±15		mV
Input offset current	$T_A = 25°C$				20	200	nA
	$T_{AMIN} \leq T_A \leq T_{AMAX}$					300	
Input bias current	$T_A = 25°C$				80	500	nA
	$T_{AMIN} \leq T_A \leq T_{AMAX}$					0.8	µA
Input resistance	$T_A = 25°C$, $V_S = \pm20\ V$			0.3	2		MΩ
Input voltage range	$T_A = 25°C$			±12	±13		V
Large signal voltage gain	$V_S = \pm15\ V$, $V_O = \pm10\ V$, $R_L \geq 2\ k\Omega$	$T_A = 25°C$		20	200		V/mV
		$T_{AMIN} \leq T_A \leq T_{AMAX}$		15			
Output voltage swing	$V_S = \pm15\ V$	$R_L \geq 10\ k\Omega$		±12	±14		V
		$R_L \geq 2\ k\Omega$		±10	±13		
Output short circuit current	$T_A = 25°C$				25		mA
Common-mode rejection ratio	$R_S \leq 10\ k\Omega$, $V_{CM} = \pm12\ V$, $T_{AMIN} \leq T_A \leq T_{AMAX}$			70	90		dB
Supply voltage rejection ratio	$V_S = \pm20\ V$ to $V_S = \pm5\ V$, $R_S \leq 10\ \Omega$, $T_{AMIN} \leq T_A \leq T_{AMAX}$			77	96		dB
Transient response	Rise time	$T_A = 25°C$, Unity Gain			0.3		µs
	Overshoot				5%		
Slew rate	$T_A = 25°C$, Unity Gain				0.5		V/µs
Supply current	$T_A = 25°C$				1.7	2.8	mA
Power consumption	$V_S = \pm15\ V$, $T_A = 25°C$				50	85	mW

Fig. 7.12 LM741 datasheet information

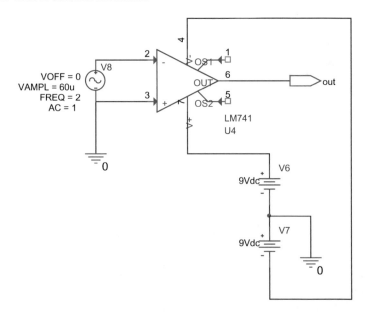

Fig. 7.13 LM741 differential gain evaluation with Bode analysis

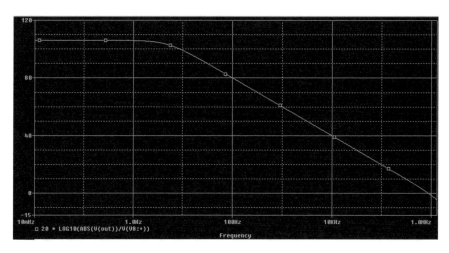

Fig. 7.14 LM741 Bode analysis

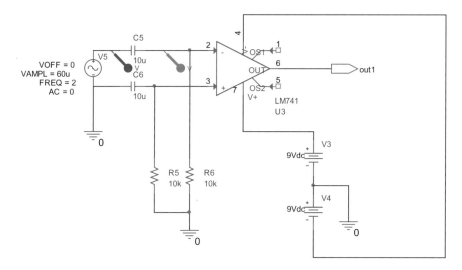

Fig. 7.15 LM741 with a differential input

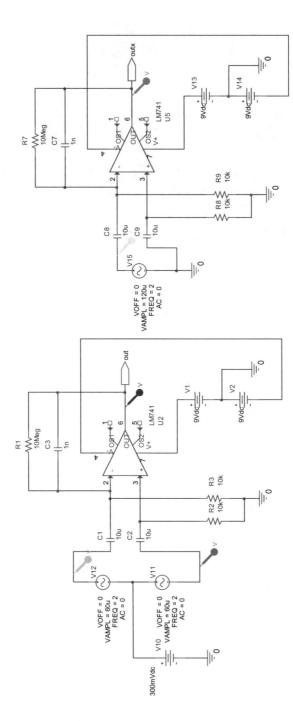

Fig. 7.16 Analysis with DC noise and with differential signals

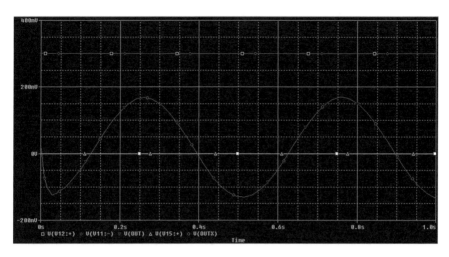

Fig. 7.17 Signals probe out and out*x*

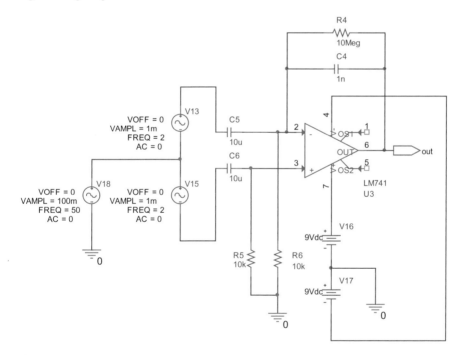

Fig. 7.18 Noise generator (V_{18}) like a common mode

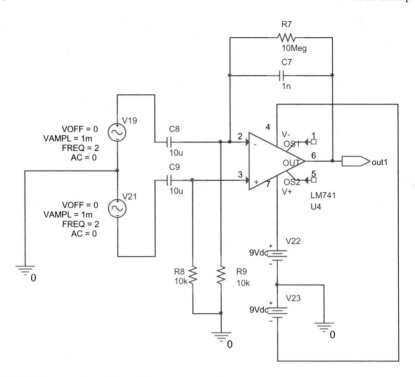

Fig. 7.19 Same configuration without common noise

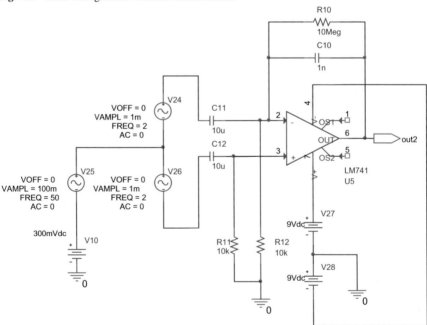

Fig. 7.20 Same configuration with common DC and AC noise on the input

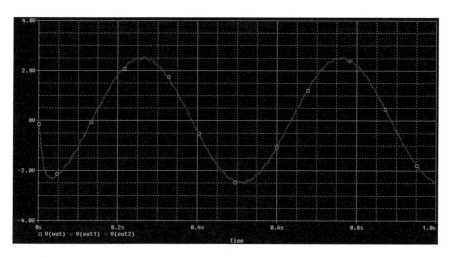

Fig. 7.21 Simulation for the three outputs, V_{out}, V_{out1}, V_{out2}

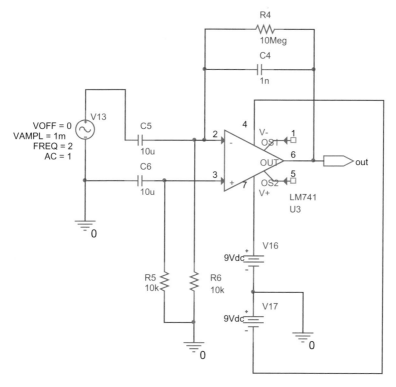

Fig. 7.22 Circuit reduction for Bode analysis (Fig. 7.19)

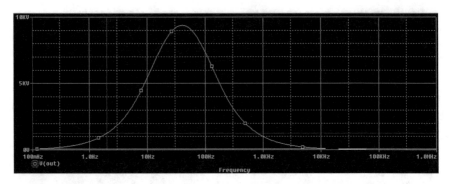

Fig. 7.23 Bode analysis for Fig. 7.21

Verify the maximum ratio between signal and common disturb can be accepted. Do an hypothesis to decide the maximum value of the AC common noise.

Analyze the maximum value of the disturb on the supply voltage using the PSRR concept.

Define the tolerance acceptable for the different passive components.

Define the best input impedance of the Op Amp.

What about the feedback network done with the R_4 (10 Meg) and C_4 (1 n)? This is a filter to stop high frequency noise that could be not equal for the two inputs. The feedback network decrease the gain over few tens of Hz. Try to study what happen changing this feedback network component values.

Write an analytic expression to find the pole and zero equation.

7.4   Radio Transmitter

Now we want to analyze an FM transmitter. Reading many of the books or electronic magazines you can find many examples of receivers or transmitters in many different configurations. We choose one of the circuits we found and we would like to show how to simulate and understand it.

I think that the most part of the analog electronic is contained in this circuit; all the theory students had studied during many years on the school desks are summarized in this circuit, as you can see immediately.

In the first stage, we found is a low-frequency amplifier, to amplify the signal coming from the microphone. This amplifier is composed by an operational amplifier with gain and filtering capability. The output of this amplifier is a voltage function of the input signal.

This voltage will be used to vary the capacitance of a VARICAP device, in the second stage, a solid-state capacitor able to vary its capacitance depending from the voltage applied to generate a change in oscillation of an RF oscillator. Finally, the

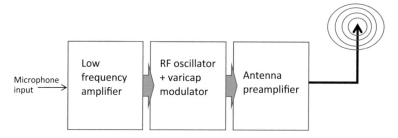

Fig. 7.24 Three stages composing the RF transmitter

third stage is another amplifier to drive the signal antenna with the right impedance, as shown in Fig. 7.24.

Before to show the complete circuit, we will study the three stages separately. Assembly all together, and turn on the supply, could be easier, but we want to understand. If you want, in the future, to design circuits, you have to understand the behavior of each part. Read a schematic is like to read a musical score, you read and translate in your mind, the music. For a circuit, you have to read and imagine to be the electromagnetic wave and the potential, forgetting the schematic, see only the behavior of the electrons.

Ok, let start with the intermediate stage, the RF oscillator with the VARICAP. Figure 7.25 shows the RF oscillator; this is not the complete schematic we need, and the VARICAP is not yet present.

Our first analysis will be concentrate on the RF oscillator, a Hartley configuration.

This type of oscillator has three reactive elements, and in this case, two inductors, L_{16} and L_{17}, and one capacitor, C_2, and one amplifier component, the transistor Q_1, have a very detailed story and the analysis, referred to all the parameters we want to considerate, is well done.

Probably it is better to spend few words to recall how the LC configuration, normally called 'tank,' meaning that we have in that the energy reserve needs to produce the oscillation, works, as shown in Fig. 7.26.

It is important to said that the reactive component must have the initial condition at time $= 0$ for a meaningful simulation (.ic $v(x) = 0$).

The simulation shows the oscillations produced by the dribble of the energy between the C and the L. The results, in this case, are a dumped oscillation because the tank is not continuously stocked with new energy. C and L are connected to the power supply, but the system will reach a stable condition, after a proper time, because I can say, the energy is not 'properly managed'. Figure 7.27 shows the currents in the upper window and the voltages in the lower window.

After turn on, the DC point is around 6 V because the voltage divider resistor, done with R_{23} and R_{22}, is around 2/3; the capacitor will start to charge, red probe point voltage increases and the same happen for the inductors, blue probe point. Current in the capacitor increases (green) and in the inductor increases again but in the opposite direction. Current in R_{23} decreases till the capacitor is charged. After

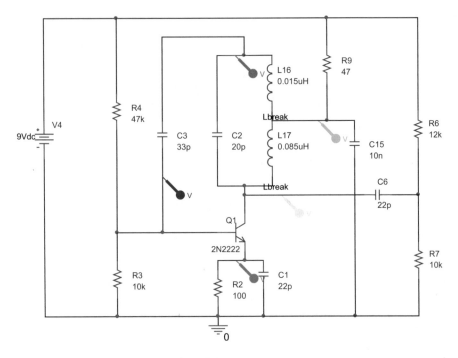

Fig. 7.25 Hartley oscillator for the RF module

Fig. 7.26 LC tank

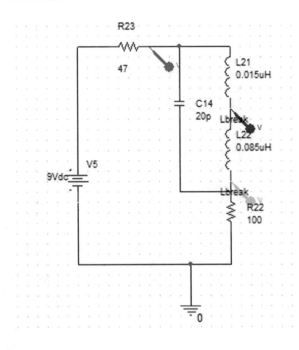

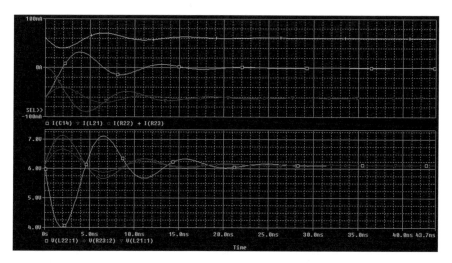

Fig. 7.27 LC tank simulation

2 ns, in Fig. 7.27, the voltages on the capacitor reaches the maximum value and, in this moment, the current capacitor is 0. The capacitor is charged and no more currents flow to charge it. Now you have to remember, with an engineering way to speak that the capacitor during the charge action has current and voltage at its extremities, with opposite polarity, when it starts to discharge the voltage remains with same polarity had during the charge but the current turns its verse. Vice versa, for an inductor, during the charge the current and the voltage have opposite polarities too, but when the charge is complete, the electromagnetic field collapses immediately and current remains in the same verse, but the voltage, at the inductor extremities, is reversed. C and L are components with complementary functionality. So, we were with the C charged. It starts to discharge, read probe voltage decrease and current, $I(C_{14})$, became greater then 0, so the sign changes, meaning that current is not more absorbed by the capacitor but delivered. It is important to note that green and blue probe voltage are always opposite in sign. So, the capacitor continues to deliver current for the inductor, and again the capacitor current reaches the 0 value. We reach the symmetrical situation with respect to the turn on instant and all proceed in the same way except to the fact that the amplitude of the oscillation decreases until it disappears (of course, the energy could not disappear, we have always to take in account not only capacitor and inductance, but also resistance that dissipate the energy. This will stop the oscillation by dissipation). We have to underline that we consider two separate inductors, but we will neglect the interaction between them (mutual inductance). We can write:

$$L = L_{21} + L_{22}$$

To be more accurate, we should take into account the mutual interaction, i.e., the measure that the magnetic flow, generated by one inductor, interacts with the other inductor changing the final total inductance value, adding an M to the previous equation.

From the simulation, the period of the oscillation is ~9 ns. From the equation, we know that the frequency oscillation, when the capacitive reactance is equal to the inductive reactance, is:

$$f_r = \frac{1}{2\pi \sqrt{L * C_{14}}}$$

In our case: $f_r = \frac{1}{2\pi \sqrt{(0.015 \ \mu H + 0.085 \ \mu H)20 \ pF}} = \frac{1}{2\pi \sqrt{2 \times 10^{-6} \times 10^{-12}}} = \sim 112$ MHz.

And we know that 9 ns corresponds to $\rightarrow \frac{1}{9 \times 10^{-9}} \sim 111$ MHz.

The circuit moves the energy from capacitor to inductor, reversing the voltage every time, and producing an oscillation, we know that capacitor and inductor work only when a transition occurs; if we are not able to repeat a transition periodically, the energy, at each cycle, decreases, due to the Joule effect and electromagnetic propagation effect (inductor is always an antenna), producing a dumped oscillation. To have a continuous oscillation, with this type of circuit, we need to have two different oscillations in opposite phase, and for this reason, we have two inductances and feedback one of these oscillations to an active element, Q_1. In a real circuit, we have only one inductor with a connection on a coil position, to save space.

I suggest, before to proceed, to study the circuit writing on a paper how the simulation works and after compare your drawing with the simulation. This circuit will be never understood enough.

And now come back on Fig. 7.25. Circuit tank is what we studied. Then, we have a bipolar transistor Q_1, with its bias resistors, R_4 and R_3. Q_1 base is connected with the tank, L_{16}, L_{17} and C_2 in Fig. 7.25, by a capacitor to stop the DC current and has only the AC signal. Tank circuit is supplied by the R_9, and Q_1 has a feedback emitter circuit R_2 and C_1. C_6 and R_6, R_7 are parts of the antenna amplifier, the output of this stage. R_9 and C_{15} delay the charge for the tank supply.

How this circuit works? First, R_4, R_3 bias the Q_1 base to 1 V and keeps Q_1 on. For the Kirchhoff network equation, with a VBE equal to 0.6 V we have 0.4 V on R_2. Then, when the supply voltage is turn on, the capacitor C_2 will charge and the oscillation between C_2 and L_{16}, L_{17} will start. With C_3, the oscillation is also used to drive Q_1 base. When oscillation increases, Q_1 base voltage will increase and so the current flowing in Q_1 emitter is increasing the collector voltage. Figure 7.28 is a simulation of Fig. 7.25 circuit. We can see that the circuit reaches a stable point of oscillation, it evolves.

Yellow curve is the output, and its excursions reach two times the value of the supply voltage. Green probe voltage, the common point of the inductances, starts lower than the supply voltage because the capacitance must be charged, but, during the time evolution, green voltage reaches 9 V, the supply voltage.

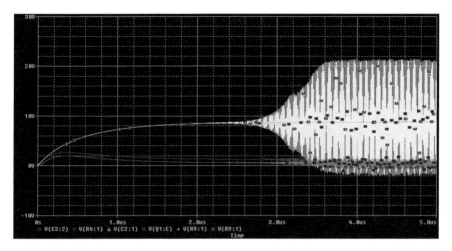

Fig. 7.28 Hartley oscillator for the RF module first simulation

Take in account to put the capacitors and inductors initial conditions set to 0 V. Pink and blue part start from 0 V and, thanks to the C_3 coupling, their values follow the tank circuit. The oscillation proceeds and increases its amplitude because the yellow voltage can increase lowering Q_1 (VCE) to 0. When Q_1 (VCE) will be lower as much as possible, not 0 (of course!), the emitter current will be at minimum value and Q_1 will be in saturation. So the minimum voltage value of the oscillation voltage will be roughly 0 V. The maximum value is near to 20 V, i.e., two times the value of the supply voltage.

Green voltage oscillates around the 9 V, the supply voltage value.

Try to summarize what happened when the circuit is at its operating point.

Figure 7.29 is a snapshot of the tank circuit, voltages and currents.

We extract the part of the schematic from Fig. 7.25, trying to simplifying the development of the voltages and current in time.

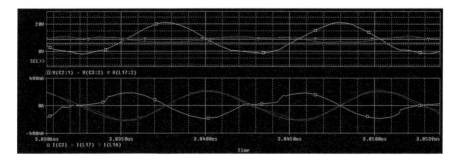

Fig. 7.29 Tank circuit analysis

Fig. 7.30 Tank circuit
analysis part 1

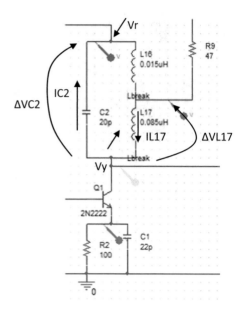

Starting from the cross point time of the currents, in Fig. 7.29 where we have the cursor, IC_2 is increasing and yellow probe is lower than red point, ΔVC_2 ($\Delta VC_2 > 0$). Yellow probe is rising too, yellow probe is always positive and IL_{17} is increasing in value, but the direction is reverse respect IC_2 (it is negative), while ΔVL_{17} is decreasing with the same sign like ΔVC_2. In this case, the capacitor is discharging, and we know this because the red voltage probe is decreasing and the inductor is charging, as shown in Fig. 7.30.

Now we are between the current's cross point and the voltage's cross points.

IC_2 increases again, decreasing the ΔVC_2 because red point lowers itself voltage. IL_{17} increases, in value, with the same direction and the ΔVL_{17}, like the ΔVC_2, decrease reaches to 0. Red, green and yellow probe voltage points are not 0 V, but around to 9 V, the supply voltage, because, at steady state, this is the voltage to be reached, as shown in Figs. 7.31 and 7.32.

Figure 7.32 Situation at the voltage cross point time, $\Delta VC_2 = \Delta VL_{17} = 0$ V.

So, the capacitor has 0 V drop across it, this means that it has no more charge inside. On the contrary, the inductance still has 0 V on its terminals, but it was loaded by electromagnetic field.

The next step is to describe the reverse situation. We had described the discharge of the capacitor and the charge of the inductor and now the reverse situation, Fig. 7.33.

We have, seeing the simulation, the current in L_{17} which is decreasing, but the verse remains always positive. The ΔVC_2 and ΔVL_{17} are reversed their polarity, means that L_{17} is discharging and C_2 is charging, with the opposite polarity with respect to the previous cycle. Voltage probe yellow now is increasing. The situation goes on until the currents, IC_2 and IL_{17}, will be equal to 0 when the ΔVC_2 and

Fig. 7.31 Tank circuit analysis part 2

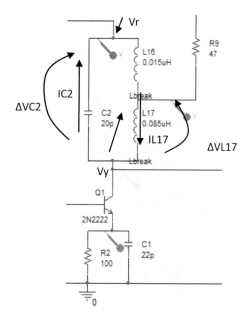

Fig. 7.32 Tank circuit analysis part 3

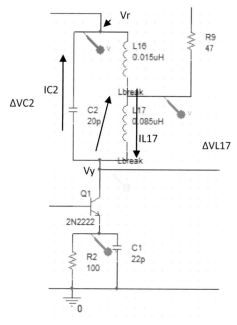

Fig. 7.33 Tank circuit
analysis part 4

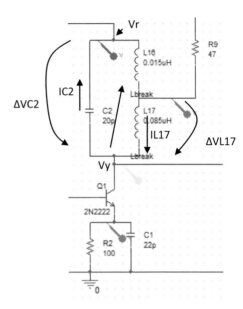

ΔVL_{17} reach their maximum value and the cycle repeats itself forever. Please read
again these last pages writing on a paper first and then running the simulations.

Now it is time to see how the first stage, the low-frequency amplifier, works.

Figure 7.34 is the low-frequency amplifier schematic. It is an operational ampli-

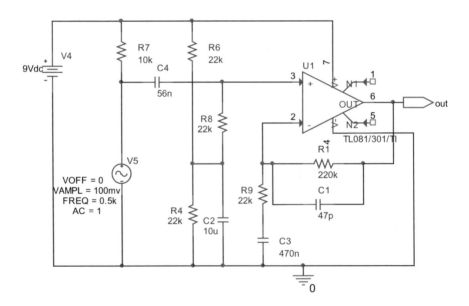

Fig. 7.34 Low-frequency amplifier schematic

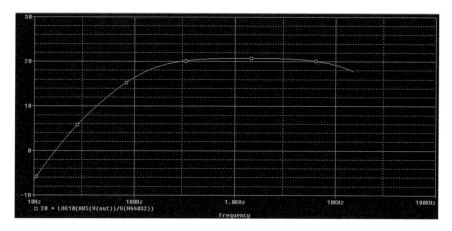

Fig. 7.35 Gain analysis for the low-frequency amplifier

fier, connected as non-inverting one. The input, V_5, represents a microphone that transmits voice to the circuit to have the voltage conversion. The output signal drives a 'VARICAP,' not yet designed in this circuit, a diode used to have a variable capacitance depending from the voltage applied. This component will be inserted in parallel in the LC tank obtaining a frequency modulation.

Figure 7.35 shows the gain analysis done, showing a gain of approximately 20 dB in the range of 300 Hz to 10 kHz. The gain is obtained by

$$\frac{R_1}{R_9} = \frac{220 \text{ k}\Omega}{22 \text{ k}\Omega} = 10$$

Figure 7.36 is the same circuit but with a PWL generator as the input, to study how the circuit works in the time domain. The result is shown in Fig. 7.37. With 0 V in input, the output is at 4.5 V and changes, with the gain, already shown.

The output goes from 4.5 to 5 V maximum. Now we have to use this voltage variation to have a frequency variation, as we said, using a VARICAP. Consider to use a BB417 as VARICAP diode, let we analyze the variation we can have on oscillation done by tank circuit.

Figure 7.38 is our oscillator circuit where we have also put the VARICAP. We change the capacitor in parallel to the inductors using a 15 pF because, now, we have the VARICAP in parallel. The final circuit needs also a compensator (a variable capacitor) because the components have, of course, some tolerances and we have to adjust, as usual, the final version circuit to compensate them.

Original circuit asked to use a BB909 VARICAP, but, as sometime happens, we don't have it in our laboratory; so, we have to use another component (in reality, we haven't the model to use to simulate) and anyway, we decided to use the BB417. Of course, we have to take in account the differences between the two diodes.

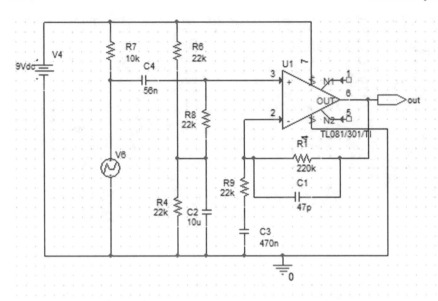

Fig. 7.36 Low-frequency amplifier with input done by V_6, a PWL generator

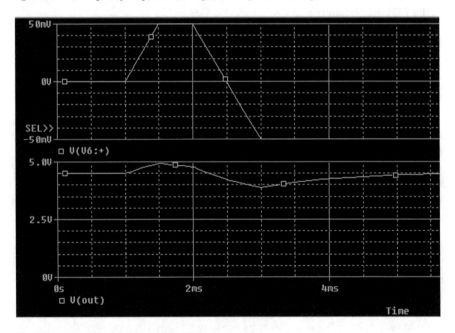

Fig. 7.37 Time-domain analysis with PWL generator

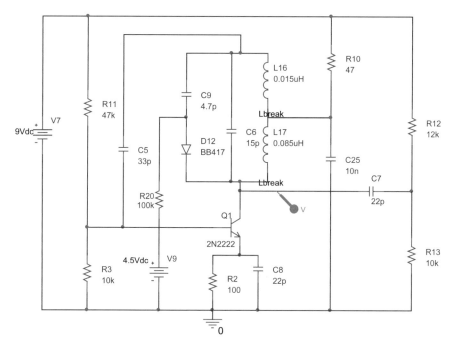

Fig. 7.38 Oscillator with the BB417 VARICAP inserted

Search the two VARICAP datasheets, and compare the values of the capacitances we obtained.

In Fig. 7.38, the BB417 is driven by a generator (V_9), a static supply voltage, for this moment, that represents the output of the low-frequency amplifier studied before. Remember that BB417 changes its capacitance value depending on the voltage applied. To insert the VARICAP in the circuit, we cannot connect directly the BB417 in parallel without interfere the tank with the DC generator, so, also for the final solution, we use a capacitor, C_9 in Fig. 7.38, to uncouple them.

Using the BB417, we have a different frequency oscillations increasing the range of the output for the low-frequency amplifier. With the BB909, the estimation was 1 V around the 4.5 V to have a change in the frequency, with the BB417, to have the change we need to move the voltage driving from 2 to 7 V. The table summarized the three simulations

	Period measured at 10 V (μs)	Frequency (MHz)
$V_9 = 2$ V	0.0094	106
$V_9 = 4.5$ V	0.0095	105
$V_9 = 7$ V	0.0096	104

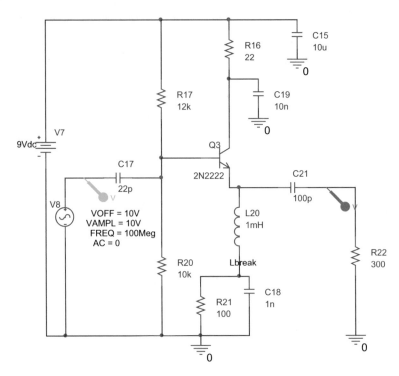

Fig. 7.39 Antenna preamplifier

We need to change the gain of the frequency amplifier studied in Fig. 7.36 to obtain the range needed. So we need to gain roughly 15 instead of 10. We can simple change a little bit the value of the resistor R_1, in Fig. 7.36, from 220 to 330 kΩ.

We have now to analyze the third stage of Fig. 7.24, the antenna preamplifier. Figure 7.39 is the circuit for the antenna preamplifier.

The circuit is, by one transistor, driven by the output of the Hartley oscillator, and in Fig. 7.39, we put a generator reproducing what we obtained from the Hartley oscillator study: 20 V of oscillation peak to peak, from 0 to 20 V and a frequency of 100 MHz.

The antenna is represented by a resistor R_{22}, 300 Ω value. The L_{20} is a RF inductance use to adapt the output of the transistor to the antenna impedance. The simulation results are represented in Fig. 7.40. After some microsecond to reach the right value of nodes, the oscillation starts. Remember to use the initial condition for the reactive element, otherwise the convergence for the simulator could be not reached.

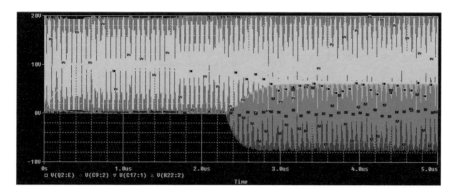

Fig. 7.40 Antenna preamplifier simulation

Try to change the value of the L_{20} and R_{22} and understand what happen. Find a model for the antenna to substitute R_{22} with a more realistic component and do the simulation.

Last step will be to put the three circuits together and try to simulate it. Of course, we cannot think to simulate a real voice exchange because the frequency of the sound is in the range of kHz that means millisecond, while the Hartley carrier wave frequency is in the MHz range. One second of real voice message means a lot of calculations and a lot of data to be represented. We have to settle for it and do only a dedicated simulation.

Figure 7.41 is the complete circuit, we add a new component, C_{22}, 10 nF, in the connection point of the two impedances and the role of this capacitance is to delay the rise of the different tank nodes. We connect the three different parts already studied and run a simulation using a VPWL generator V_6 with a sequence (0 0 10u 0 50u 0.5). We would like to see that the frequency value changes from the $V_6 = 0$ to $V_6 = 0.5$ V that correspond to a voice signal.

Simulation global results are shown in Fig. 7.42.

On this simulation results, we did a zoom before the 10 μs and close to 50 μs.

Figure 7.43 shows the oscillation closed to 10 μs, and the frequency calculated is ~111 MHz, while in Fig. 7.44, we can see a frequency of 100 MHz at 50 μs.

We can close this exercise, but, for many reasons, a lot of questions are still opened, and we suggest to study:

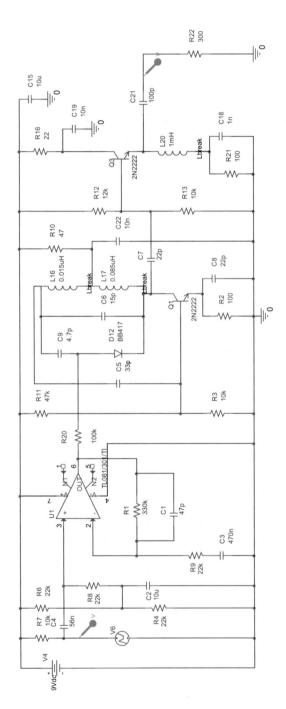

Fig. 7.41 Complete circuit for the FM transmitter

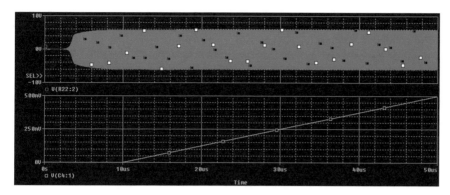

Fig. 7.42 Complete circuit for the FM transmitter simulation. Green = input, red = output as the probe in Fig. 7.25

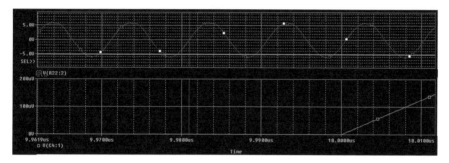

Fig. 7.43 Complete circuit for the FM transmitter simulation. Green = input, red = output. Frequency = 111 MHz

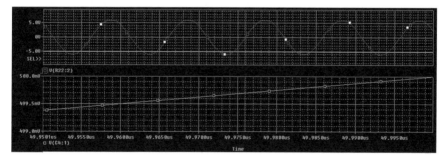

Fig. 7.44 Complete circuit for the FM transmitter simulation. Green = input, red = output. Frequency = 100 MHz

What is the power consumption of this circuit? We use a battery, how long will work with a 9 V battery?

What is the behavior of the circuit with the temperature? This must be work at −10° centigrade degree and also at 40° centigrade degree, winter and summer.

What is the window working respect the component tolerances? We have to identify the worst components, with respect to the variability and to study them.

Chapter 8
Simulation of Device Characteristics

In this chapter, we work on the device characteristics, the identity card for a device. As we did in the other chapters, the topic will be a chance to see also different concepts. Characteristics, V–I graph, are fundamental also for a complex circuit, done by lot components, as we will discover.

8.1 Introduction

What are the device characteristics? What does it mean to measure a characteristic? Normally, we refer these as V–I characteristics stand for voltage–current characteristics. The V–I graph yields valuable information about the resistance and breaks down an electronic component. It also provides the operating region of a component, and by studying these characteristics, we can understand where and how to use a component in an electronic circuit, and component is not only a discrete component but could be a circuit, a board. In V–I characteristics, the voltage, V, is normally on the x-axis, and the current, I, is on the y-axis because it is easier to control the applied voltage rather than current. This makes the voltage the independent variable and is hence traditionally placed on the x-axis. In this chapter, we will try to explore how to simulate the characteristic, using the model we have in our simulator, taking in account that you can find the same information in the datasheet. Moreover, we will investigate how to build circuit to measure, in laboratory, the same characteristics.

As for a child, when he learns to speak, the first words must be, when one thinks of electronics, are voltage and current. When I started to study electronics, it was first electrical engineering, all the teachers said that all the electronics could be summarized with few elements: R, L and C, voltage and current generator, and Thevenin and Norton theorems are the most important 'keys' to open all your mental blocks. Now I understand.

Every electrical phenomenon can be described in some form based on these two fundamental properties. Despite this, much of electrical engineering devolves away from these fundamentals and understanding is obscured by adding layers upon the

© Springer Nature Switzerland AG 2020
R. Gastaldi and G. Campardo, *Electronic Experiences in a Virtual Lab*,
https://doi.org/10.1007/978-3-030-45179-0_8

foundation of voltage and current. The question that needs to be asked of any electronic component is how the voltage is related to the current. This understanding brings powerful insight into the design of electronic circuits.

We will study, not necessarily in this order: resistor, diode, transistor, BJT and MOS, operational amplifier, showing their electric characteristics and some related circuits, not only to measure pure V–I single component characteristic but also to show how to measure electrical characteristics generally speaking.

8.2   Resistor and Impedance

Let me start with the resistor.

Simulation shows the V–I characteristic for a resistor as shown in Fig. 8.1. Easy! If the temperature changes, the value of the resistor changes. R_2 has zero for the temperature coefficient variation. In R_1, we put a coefficient equal to 0.004041, the α value for the copper. The relationship for the resistance value versus the temperature is

$$R = R_{\text{ref}}[1 + \alpha(T - T_{\text{ref}})]$$

where R_{ref} is the reference temperature, equal to 27° centigrade degree.

The two simulations in Fig. 8.1 are the characteristic at three different temperatures: 0, 27 and 125 °C. The first is the R_2 three characteristics, and they are exactly the same. The second simulation shows the differences, with the temperature.

Temperature variation is one of the most important parameters to take in account during the design. For example, to design a circuit for an automotive application, with a junction, temperature ranges from −40 to 165 °C which could be really a challenge.

If the electrical supply voltage, V, applied to the terminals of the resistive element R above was varied, and the resulting current, I measured, this current would be characterized as: $I = V/R$, being one of Ohm's law equations. The V–I characteristic curves define the resistive element, in the sense that if we apply any voltage value to the resistive element, the resulting current is directly obtainable from the V–I characteristics. As a result, the power dissipated (or generated) by the resistive element can also be determined from the V–I curve.

If the voltage and current are positive in nature, then the V–I characteristic curves will be positive in quadrant I, and if the voltage and therefore the current are negative in nature, then the curve will be displayed in quadrant III. In a pure resistance, the relationship between voltage and current is linear and constant at a constant temperature, such that the current is proportional to the potential difference V times the constant of proportionality $1/R$. Then, the current through the resistor is a function of the applied voltage, and we can demonstrate this visually using a V–I characteristic curve.

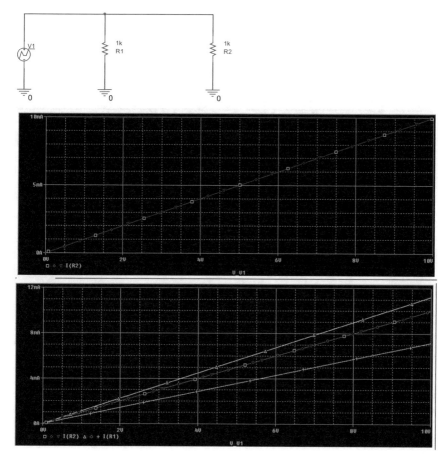

Fig. 8.1 Scheme for resistors. R_2 hasn't temperature variation, R_1 has. First simulation is related to R_2, second to R_2 and R_1

In this simple example, the current I against the potential difference V is a straight line with constant slope $1/R$ as the relation is linear and ohmic. However, practical resistors may exhibit nonlinear behavior under certain conditions for example, when exposed to high temperatures. Figure 8.2 shows a NTC characteristic, means Negative Temperature Coefficient, in temperature and the related schematic.

An NTC is a resistance with a negative variation versus the temperature, while a normal resistance has a positive one. NTC could be used in feedback circuit.

We can have also the necessity to measure other types of passive component, like the impedance.

If we connect an amplifier to a speaker, a sound box and I don't know the impedance of the speaker, if the impedance is too high, respect the amplifier output impedance we will have not the performance we suppose. Vice versa, if the speaker

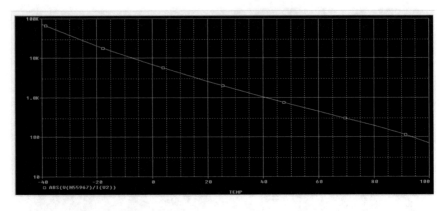

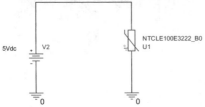

Fig. 8.2 NTC characteristic and the scheme, only to show what type of component we used for the simulation

impedance was too low, we can burn the final-stage transistors due to the overloading current.

An example could be: We have an amplifier designed to have 30 W for a 4 Ω load. What happen if we connected the amplifier with an 8 Ω impedance?

Working with the root mean square (R.M.S.), we can simply find

$$V_{\text{eff}} = \sqrt{P \times Z} = \sqrt{30 \times 4} = 10.95 \text{ V}$$

$$I_{\text{eff}} = \sqrt{\frac{P}{Z}} = \sqrt{\frac{30}{4}} = 2.73 \text{ A}$$

where P is the power and Z the impedance.

The result is that our amplifier is able to provide, at maximum, 10.95 V with 2.73 A.

If we connect to the amplifier, a speaker with 6 Ω impedance, instead 4 Ω, as required by design, we will have that the maximum R.M.S. voltage will remain the same because this value is done by the supply voltage generator but the power decreases. If we connect the amplifier with a lower speaker impedance, with the hypothesis that the amplifier is able to deliver all the current we required maintaining the output voltage, the power, for the speaker, will be greater with a possibility to damage it.

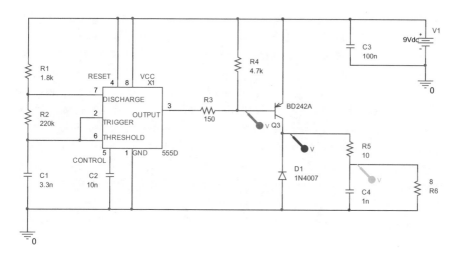

Fig. 8.3 Circuit to generate a square wave able to drive 0.5 A on 8 Ω resistive load

It is simple to understand that we cannot measure this impedance with a DC voltmeter because the impedance is a function of the frequency.[1]

Normally, the speaker impedance is measured at 1 kHz. To do this in the right way, we have to use a sinusoidal generator, but to simplify the project we can use a square wave generator, as shown in Fig. 8.3.

A first simulation is shown in Fig. 8.4, to see the voltage node and the current sunk by the R_6 load.

Of course, the impedance, for a loudspeaker, is a value depending from the frequency; manufacturer calculates the impedance at 1 kHz. Out circuit doesn't generate a sinusoidal wave; this will be the right thing to do, but a simple square wave.

Can you calculate the error introduced using a square wave instead a sinusoidal one?

We use an NE555, a well known integrated used as a timer driving a power BJT, to have, on the output load, at least 0.5 A.

Figure 8.5 is the NE555 block schematic from the relative datasheet. You can find in the DS all the explanation useful to understand and use the integrated circuit. I suggest to simulate the different configurations you can find.

Coming again on our circuit, a possible circuit to simulate loudspeaker impedance is shown in Fig. 8.6.

Simulation to study frequency behavior gives a result, for the impedance, as shown in Fig. 8.7. Here we can see how the impedance changes versus the frequency.

As we can read in the cursor table, under the waveform, till 20 Hz, the impedance is less than 6 Ω. The frequency resonance peak is far from the usable range.

---

[1]I remember, in a company where I worked many years ago, once a boy, with a DC voltmeter he entered the warehouse and after some time he came out shouting: 'we have all the transformers shorted!'.

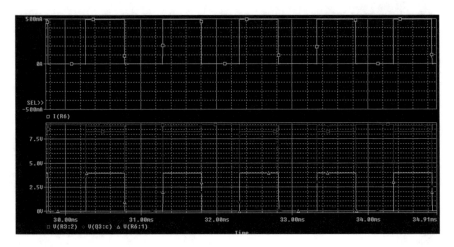

Fig. 8.4 Simulation voltage and current

Fig. 8.5 NE555 block
schematic from the DS

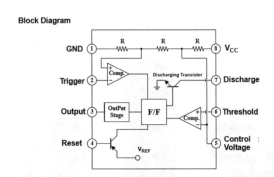

Fig. 8.6 Loudspeaker
network model

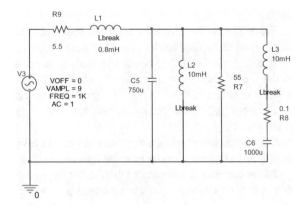

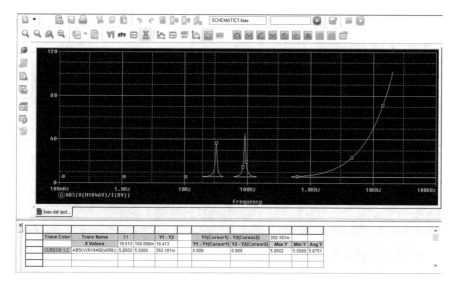

Fig. 8.7 Impedance versus frequency

So, finally, to use our circuit, we have to consider that we are using a square wave and not a sinusoidal.

Figure 8.8 shows the schematic and Fig. 8.9 the simulation.

Figure 8.9 is a zoom to show a particular point to be highlighted.

Using a square wave needs to be care with the very fast edge that we have to neglect. We have to take into account only 1 kHz frequency and not the highest frequency contained in the edge. The impedance will be the value obtained in the 'stable' part of the square wave. In the simulation, it is less than 6 Ω, again.

Another interesting possibility is to build a circuit to measure the impedance for an antenna. We know that receivers and transmitters have and impedance equal to

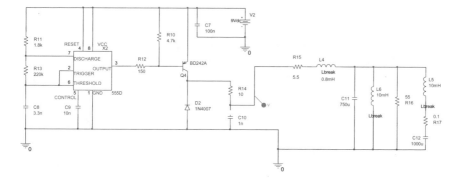

Fig. 8.8 Schematic to calculate the impedance for the circuit in Fig. 8.6

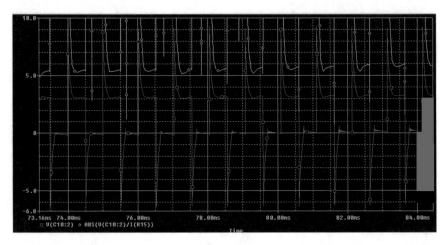

Fig. 8.9 Impedance evaluation

52 Ω as the coaxial cables. If the impedance has not this precise value, we will face a mismatching with a consequent power loss. You know that the antenna impedance depends from the assembly, from the height where you put it, for example. Having a way to measure the antenna impedance could be very interesting to better fit the device performances.

To do this, we will use a high-frequency Wheatstone bridge. With this circuit, we are able to check if the value of the impedance, at certain frequency, of the working frequency of the antenna has the right value. The schematic in Fig. 8.10 shows the circuit to be used.

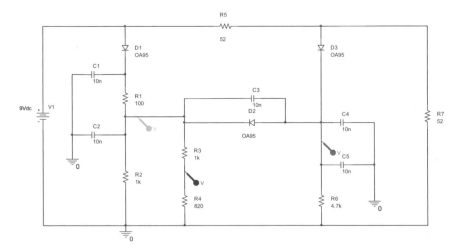

Fig. 8.10 High-frequency Wheatstone bridge

We use, for this simulation, a DC input, V_1. The antenna is represented by the R_7, 52 Ω.

The concept is very easy: A high-frequency generator will be placed instead V_1, working at the frequency for which the antenna was designed. D_1 with C_1 and C_2 will rectify the current from V_1 giving a DC voltage at the blue probe. R_3 must be a potentiometer, and the blue point must be withdrawn from it. The same for the red probe, where the DC voltage will be function of the R_7 value. Placing a voltmeter between blue and red probes, it is possible to verify when the voltage is zero and then obtain the value of the antenna impedance.

8.3 Diode

The diode is the first semiconductor device normally studied. Like transistors, thyristors are all constructed using semiconductor PN junctions connected together and as such their $V–I$ characteristics curves will reflect the operation of these PN junctions. Then, these devices will have nonlinear $I–V$ characteristics, as opposed to resistors which have a linear relationship between the current and voltage.

So, for example, the primary function of a semiconductor diode is rectification of AC to DC. When a diode is forward biased (the higher potential is connected to its anode), it will pass current. When the diode is reverse biased (the higher potential is connected to its cathode), the current is blocked. Then, a PN junction needs a bias voltage of a certain polarity and amplitude for current to flow. This bias voltage also controls the resistance of the junction and therefore the flow of current through it.

When the diode is forward biased, anode positive with respect to the cathode, a forward, or positive current, passes through the diode and operates in the top right quadrant of its $V–I$ characteristic curves. Starting at the zero intersection, the curve increases gradually into the forward quadrant, but the forward current and voltage are extremely small.

When the forward voltage exceeds the diodes PN junction's internal barrier voltage, which for silicon is about 0.7 V, the forward current increases rapidly for a very small increase in voltage producing a nonlinear curve. The 'knee' point is on the forward curve.

Likewise, when the diode is reverse biased, cathode is positive with respect to the anode, the diode blocks current except for an extremely small leakage current, and it operates in the lower left quadrant of its $V–I$ characteristic curves. The diode continues to block current flow through it until the reverse voltage across the diode becomes greater than its breakdown voltage point resulting in a sudden increase in reverse current producing a fairly straight line downward curve as the voltage loss control. This reverse breakdown voltage point is used to good effect with Zener diodes.

Then, we can see that the $V–I$ characteristic curves for a silicon diode are nonlinear and very different to that of the previous resistors linear $V–I$ curves as their electrical characteristics are different.

The trick, for a diode, is the depletion region formed at the time of manufacturing, where electrons and hole combine to form ions and no carrier is available for conduction in the region. N-region has free electrons as majority carriers and P-region has holes. During the absence of an external voltage source, because of the random movement, minority carriers can enter the depletion region. Any hole entered to the depletion region will be attracted to P-region, and the free electron will be attracted to N-region by the attraction force of opposite ions in the depletion region. Some minority carrier may enter the P-type, and others may enter the N-type and there will be no net current flow.

In reverse bias, the anode terminal of the voltage source is connected to the N-type pin and cathode terminal of the voltage source is connected to the P-type pin of the diode. In the reverse bias operation, the diode acts like an open switch. The anode terminal of the source will draw the free electrons from N-type, and cathode will draw hole from P-type. Thus, the number of ions in N-region and P-region will increase which is the reason for the widening of the depletion region.

However, the minority carrier will enter the depletion region and pass to the other side of the junction causing a small current. This small amount of current is called reverse saturation current. The term 'saturation' shows the fact that after a very short change in the current for a change in voltage, the current will not increase any more for increasing the reverse bias voltage.

In the V–I characteristic curves, the graph in the third quadrant represents the reverse behavior of the diode. In the beginning, the current changes very quickly for a small change in voltage and reaches the saturation current. The further changes in the voltage do not affect the current.

For tens of volts in reverse bias, the current remains constant. But reaching some high reverse voltage causes a huge current in reverse direction. As the reverse voltage increases, the velocity of the reverse current increases as well as its kinetic energy. The high kinetic energy is transferred to valence shell electrons of stable atoms and makes them leave the atom. These additional carriers can aid in the reverse current flow. The voltage at which this rapid change in current occurs called Zener voltage. This is in the third quadrant of the V–I characteristic curve.

A diode is forward biased if the P-type pin is connected with the anode of a voltage source and N-type of the diode is connected with the cathode of the source. In forward bias operation, the diode acts like a closed switch. The voltage source in forward bias configuration applies pressure on free electrons in N-region and holes in P-region toward the depletion region. The free electrons and hole recombine with the ions near the depletion region, and the depletion region width is reduced. Then, the majority of carrier can pass the thin depletion region. As the forward bias voltage increase, the depletion region width decreases and more carrier can pass.

The first quadrant of the V–I characteristics curves shows the forward operation of the diode. In the beginning, by increasing the voltage, the current changes very slowly but when the voltage reaches 0.7 V (for silicon), the current starts to change rapidly for a small change. The rapid change in the current shows that the resistance decreases as we increase the voltage above the knee of the curve.

Figure 8.11 is the simple scheme to simulate a diode; we choose a 1N4001.

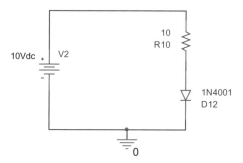

Fig. 8.11 Simulate the 1N4001 characteristic

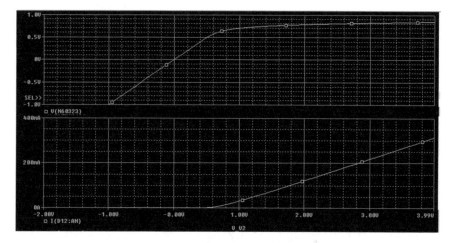

Fig. 8.12 1N4001 current, graph below and voltage across the diode, the graph above

As shown in Fig. 8.12, the simulation is done using the parametric DC sweep, at 27 °C degree.

Let we show a very simple circuit to verify if the diode you have is 'good' or not. I mean if the diode works or if it has shorted or open. If you built the circuit, pay attention to the current flowing in the diode. A critical design must be performed to verify also the power dissipation.

Anyway, the circuit, in Fig. 8.13, works like a flip-flop. R_3 and R_4 represent two lamps, and V_1 could come directly from a transformer. The two diodes, D_3 and D_4, must be selected depending on the current we plan to have in the circuit. For the simulation, we use a 'common' Dbreak.

The test diode is the D_5 as shown in Fig. 8.13. If the diodes work what happened? This depends on the biasing of D_5 in the circuit. Connected like as shown in Fig. 8.13, the results are in the simulation as shown in Fig. 8.14.

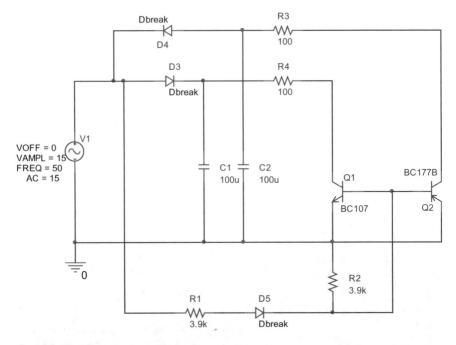

Fig. 8.13 Diode test circuit

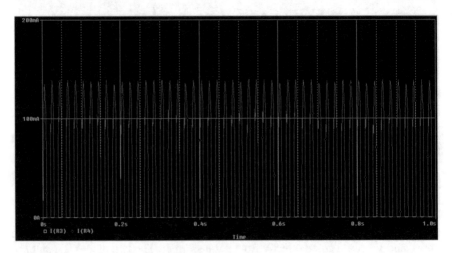

Fig. 8.14 Current result for circuit as shown in Fig. 8.13

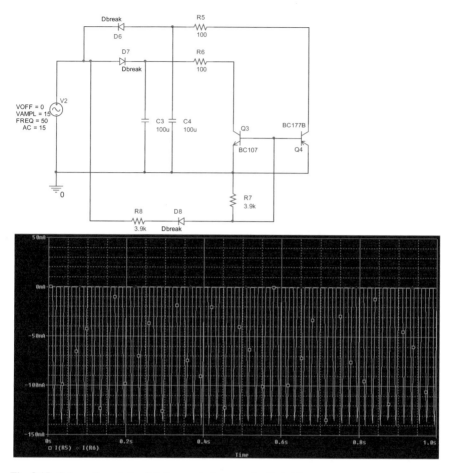

Fig. 8.15 Schematic and simulated currents for opposite diode bias

The current will flow in R_4 and not in R_3.

If we connect the diode in the reverse way, the result, schematic and simulation are as shown in Fig. 8.15.

And, what happen if the diode is not 'good,' if it is shorted?

See the schematic and the simulation in Fig. 8.16. The result is that both the lamps will be on.

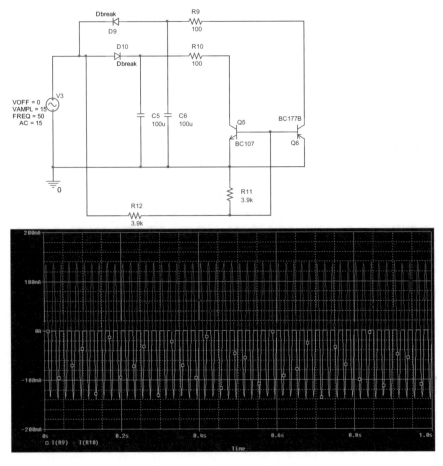

Fig. 8.16 Schematic and simulated currents for a shorted diode case

The last situation is with a broken diode, no connection, in Fig. 8.17 schematic and simulation.

Of course, in this case we have no current on both the lamps.

In Fig. 8.17, both currents are zero but we have a 'spike' at the starting point. Why?

Try to verify the power consumption for each component to design the right circuit depending on the dissipation power needed for each component.

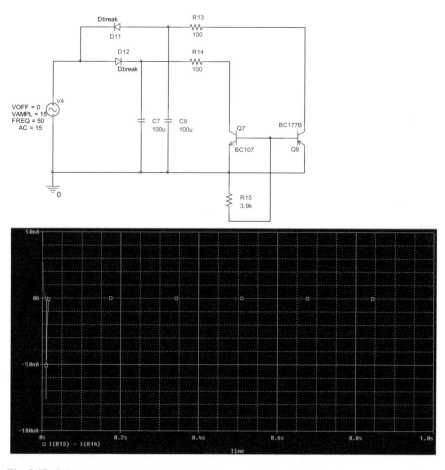

Fig. 8.17 Schematic and simulated currents for an open broken diode case

8.4 LED Diode

LED stands for Light-Emitting Diode which is used in all types of semiconductor diodes with electrical characteristics similar to a PN junction diode.

To turn on a LED, we have to bias it forward, remember also to put a series resistance to limit the current. The only difference with respect to a 'normal' diode is the threshold voltage. Threshold voltage value is strictly related to the light color emitted.

First LED was marketed during the 60 years of the last century with an infrared light of visible red color. During the following 70 years, orange, yellow and green LEDs were developed, while the shorter wavelengths (corresponding to higher frequencies) were achieved only in the 1990s.

Color	Wavelength (nm)	Threshold voltage (ΔV)
Infrared	$\lambda > 760$	$\Delta V < 1.9$
Red	$610 < \lambda < 760$	$1.63 < \Delta V < 2.03$
Orange	$590 < \lambda < 610$	$2.03 < \Delta V < 2.10$
Yellow	$570 < \lambda < 590$	$2.10 < \Delta V < 2.18$
Green	$500 < \lambda < 570$	$1.9 < \Delta V < 4.0$
Blue	$450 < \lambda < 500$	$2.48 < \Delta V < 3.7$
Violet	$400 < \lambda < 450$	$2.76 < \Delta V < 4.0$

Schematic and simulation in Fig. 8.18 show seven different diodes declared with different colors.

Check if the thresholds have the right values.

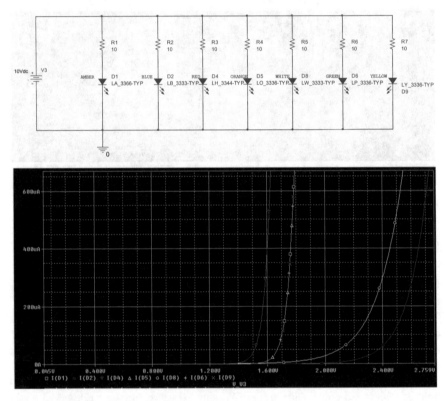

Fig. 8.18 Different LEDs, different colors

8.5 Zener Diode

Zener is a particular diode, used always in reverse biasing, to maintain constant a voltage. Normally, an integrated regulator circuit is used, but a possible solution, not so well controlled, could be a Zener.

A simple example is proposed in Fig. 8.19.

Figure 8.19 shows a simple application for a Zener diode. The two circuits are the same with a difference: R_2 and R_4 are the loads and are very different. In spite of this, the voltages, red and green probes, show the same value, Fig. 8.20.

As the datasheet said, the nominal voltage for the Zener used is 5.1 V, as the simulations showed. This voltage remains stable also if the load changes from 1 k to 500 Ω.

To obtain this value, the Zener sinks current.

An interesting point to study is the maximum variation for the load resistance to maintain the voltage required and the maximum speed admitted for this voltage to remain stable.

Do a simulation to show the V–I characteristic for this Zener diode.[2]

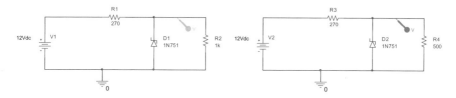

Fig. 8.19 Zener working mode with different loads

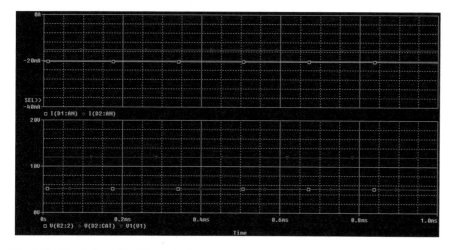

Fig. 8.20 Simulation with different loads

---

[2]Zener Theory and Design Considerations, Handbook, HBD854/D, Rev. 1, Dec-2017, ON semiconductor.

8.6 BJT and MOS Transistor Characteristics

BJT stands for Bipolar Junction Transistor, you know. In principle, it is built using three junctions, like to put two diodes in opposite connection, anode of the first one connected to the anode of the second. Unfortunately, we cannot explain the transistor functionally like two diodes because one of the principal points, to explain how the BJT works, is about the diffusion current on base and, for this reason, the base must be thin.

Anyway, the BJT is a three-terminal device, with three different fundamental assemblies, in a circuit, called common emitter, common base and common collector. The characteristic curves will be a function of the different assembly.

Moreover, the diode could be studied using two electrical quantities, whereas the BJT needs four electrical quantities.

MOS is for metal oxide semiconductor. We speak a lot about these components throughout the book. Now it is only few words to show how it could be interesting to reproduce the characteristics to appreciate the complete functionality of the device.

Consider a typical BJT, like a BC108, a general part NPN transistor.

The two main characteristics are the input and output, these must be considering for the three different configurations.

With a common emitter configuration, the output characteristic is a graphic of I_{CE} versus V_{CE} for a family of input current I_B for a scheme in Fig. 8.21 and in Fig. 8.22, how to use the simulation profile. The output characteristic graphic is shown in Fig. 8.23.

In the same way, it is possible to define, also for the common base and common collector, graphs with the characteristics.

These are the main graphs needed to define the BJT functionality we can say in DC. Of course, we need to have also other information to know the behavior in frequency and last but not least, for the power management, to avoid to burn the circuit and all it is connected. As a friend said: 'these beasts, referring to the capacitors, when they die they stink a lot.'

Fig. 8.21 To trace output
BJT characteristics

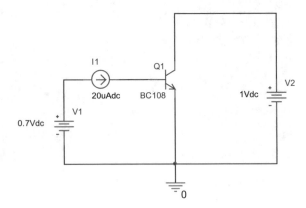

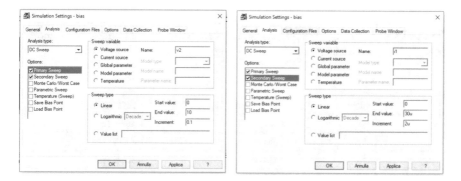

Fig. 8.22 Simulation setting to generate the characteristics

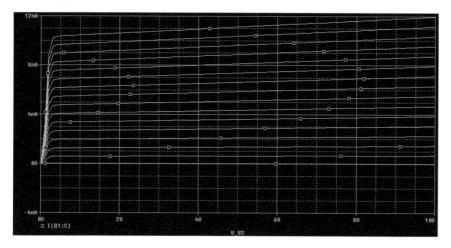

Fig. 8.23 Output characteristic curves

8.7 Operational Amplifier

We used many times, in this book, the operational amplifier. This is one of the most useful integrated circuits. The Op Amp you can find on the market works really close to the ideal component, but they aren't, of course. The Op Amp parameters are a lot, but we would like to show the main of them and how to simulate their measure, to do a different exercise with respect to the other lot of good books of the same item.

The main parameters for the real Op Amp are reported in the next table. We put the ideal value and the real one for an Op Amp done by BJT, the uA741 and one done by FET, the TL081.

	Ideal	BJT input (like uA741C)	FET input (like TL081)
R_i	∞	1–2 MΩ	10^{12} Ω
V_{os}	0	20 mV	5 mV
I_{bias}	0	0.1–5 μA	10–50 pA
I_{os}	0	20 nA	0.5–25 pA
V_{in}	+VCC, −VEE	+14 V, −14 V	+14 V, −14 V
R_o	0	50 mΩ	70 Ω
V_{out}	+VCC, −VEE	+14 V, −14 V	+14 V, −14 V
I_{sc}	–	25–90 mA	15–40 mA
A_{OL}	∞	10^5	10^5
SR	∞	0.5 V/μs	13 V/μs
GBP=f_T	∞	1 MHz	1–4 MHz
CMRR	∞	90 dB	90 dB

Let's summarize the meaning for the different parameters shown in the table.

R_i is the differential input resistance without any feedback.

Figure 8.24 shows the schematic and the simulation; we can see the value of the input current, $I(V_2)$, and in the upper plot, the calculation by Ohm's law, for the input resistance. The value is 2 MΩ.

V_{os} is the input offset voltage. This is the voltage value to be applied to have the output at 0 V.

Again, Fig. 8.25 shows the schematic and the related simulation. We have applied the same input voltage to both the inputs, an output equal to ~25 mV. In the application, we have to know if we need to take in account this shifting value. Ideally, with 0 differential input voltage, the output must be 0 too.

I_{bias} is the average value between the two input currents. Of course for the FET input Op Amp, these currents are very low, not so low for the BJT input Op Amp. The effect, due to this current, is to have the input to a certain voltage, and the output voltage will be not 0 V also with same voltages applied to the inputs.

I_{os} is the differences between the two input currents, due to the asymmetry in the differential input stage. This effect, again, has the results to generate an output voltage, different to 0 V also with the same inputs voltage applied.

V_{in} is the range for the inputs voltage and normally can reach the supply voltage values or, limited to 1–2 less than the supply voltages.

R_o is the output resistance, of course in open-loop configuration.

V_{out} is the range for the output voltage and normally can reach the supply voltage values or, limited to 1–2 less than the supply voltages.

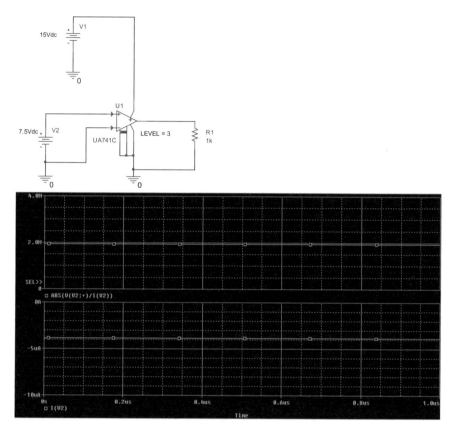

Fig. 8.24 Input resistance calculation

I_{sc} is the short-circuit current, and it is the maximum current that the Op Amp can provide.

A_{OL} is the large signal gain in an open-loop configuration. This is the intrinsic Op Amp gain, changing of course with the frequency.

SR is the slew rate. It is the maximum speed variation for the output voltage from positive to negative saturation voltages, and vice versa, with a square wave input, to have an output signal with no distortion. For a sinusoidal input, to have no distortion, it must be satisfy the relationship

$$SR = V_{OM} \cdot 2\pi f \qquad (8.1)$$

where f is the input signal frequency and V_{OM} is the maximum amplitude.

With the input equal to

$$2\pi f V_{OM} \cos(2\pi f t)$$

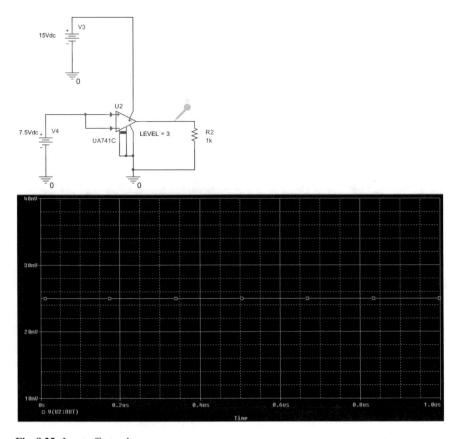

Fig. 8.25 Input offset voltage

The variation has its maximum value in the origin, where the function has the flex point and corresponds to the SR value given in (8.1).

$$\frac{dV_o}{dt} = 2\pi f\, V_{OM}\, \cos(2\pi f t) = SR$$

Do you remember where coming from the formula for the SR?
f_T is the transition frequency. The frequency at open loop, which A_{OL}, becomes equal to 1.

GBP=f_T, GBP is the **Gain Bandwidth Product, and it is constant.**

$$GBP = B \cdot A_{OL}$$

CMRR is the common-mode rejection ratio, the ratio between the differential mode gain and the common-mode gain. It is a value of the capability that does not amplify common input signals.

For all of these parameters, you can do a simulation and compare the results with table or directly to the datasheet.

8.8 Kelvin Measure

To measure a resistance, normally, we use two terminals, two wires, connected to two probes, but this system causes a large measure error if the resistance to measure is low, as shown in Fig. 8.26. R_3 represents the load we want to measure, and R_1 and R_2 are resistances due to the wire and specially the probe used for the measure. We used V_1 to represent the voltmeter and the amperometer. Of course, to know the value of the load, R_3, we impose a voltage, measure, with an amperometer in series the current and we have to use the common Ohm's law to get the R_3 value. Of course, the measure gives the total value of the series resistance. A smart way to perform the measure is to measure current and voltage in same time, like in Fig. 8.27.

Fig. 8.26 Measurement of R_3

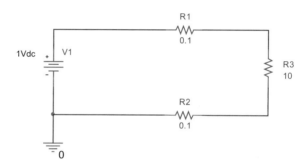

Fig. 8.27 Measurement of R_3 in a different way

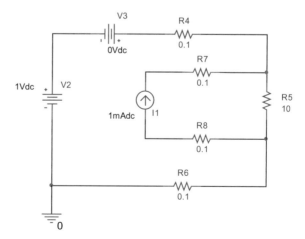

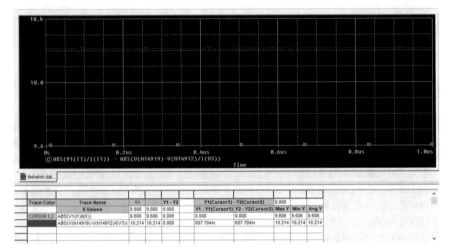

Fig. 8.28 Rload measured value

We have the same supply, V_2, to represent the real supply and another supply, V_3, of value zero, to represent the amperometer, while I_1 represents a voltmeter. Now R_5 is the load, 10 Ω; the trick is to divide the current from the voltage. I mean: Through V_3, I will measure the current flowing in R_5 because I_1 is a voltmeter, an instrument with a low output resistance able to measure the voltage to its connection. To do this, the current sunk by the voltmeter is very low, in the range of the 1 mA, so this is the same to say that the voltmeter impedance is around tens of megaohm, and this value must be compared to the resistance to be measure. Finally, the current in R_5 is equal to the current from V_3. While the voltage across R_5 is measured by with the help of I_1, with a low current does not perturb the voltage drop on R_5 due to the R_7 and R_8 wires and probe parasitic value. Figure 8.28 shows the results: Purple probe is obtained as shown in Fig. 8.28 with Ohm's law, the same but for the four points Kelvin method, the blue probe.

The error, for the Kelvin method, is the half with respect to a normal method.

This could seem to be a joke because the wish is to measure the resistance value at long distance.

This method of measurement, which avoids errors caused by wire resistance, is called the *Kelvin*, or *4-wire* method. Special connecting clips called *Kelvin clips* are made to facilitate this kind of connection across a subject resistance. This method of measure was first proposed by Lord Kelvin, in honor of him the method was called.

It is the method used to measure resistance on the boards. An invention is made in 1861! My first boss and teacher, in electronic, said to me that probably, with all of our knowledge, we will be not able to do again what the pioneers did. People in another ranking for intelligence, at least with respect to me!

Chapter 9
Virtual Prototyping Using PSpice

9.1 Introduction

OrCAD® PSpice® and Advanced Analysis Technology combine industry-leading, native analog, mixed-signal and analysis engines to deliver a complete circuit simulation and verification solution. Whether you are prototyping simple circuits, designing complex systems or validating component yield and reliability, OrCAD PSpice technology provides the best, high-performance circuit simulation to analyze and refine your circuits, components and parameters before committing to layout and fabrication. Cadence® PSpice® simulation tool performs analysis and verification of your circuits. You can use this tool to perform various analyses such as transient, AC, DC, parametric sweep and worst-case.

This chapter in this book uses circuit examples to explain the different aspects of PSpice that you can use to solve your simulation problems.

In this chapter, you will:

1. Create a simple schematic design using OrCAD® Capture.
2. Create a PSpice profile.
3. Run a simulation.
4. View simulation results in the PSpice Probe window.

You can skip this chapter if you already know the basics of Capture and PSpice. Refer to PSpice Online Help for more information.

This chapter is authored by Alok Tripathi, Product Engineering Architect, Cadence Design System Inc., alok@cadence.com. Cadence Design Systems (India) Pvt. Ltd., Plot Nos 57 A, B & C, Noida Special Economic Zone, PO NEPZ , Noida—201 305

© Springer Nature Switzerland AG 2020
R. Gastaldi and G. Campardo, *Electronic Experiences in a Virtual Lab*,
https://doi.org/10.1007/978-3-030-45179-0_9

9.2   Creating a Design

1. In Capture, choose *File → New → Project* to create a new project.

   The New Project dialog appears.

2. Specify a *Name* and *Location* for the project as shown in Fig. 9.1.
3. Select *Enable PSpice Simulation* as you will simulate the design using PSpice.
4. Click *OK*.

   The Create PSpice Project dialog appears.

5. Select *Create a blank project* as shown in Fig. 9.2 and click *OK*.

Note: In Capture, you can either create a blank project or base the new project on an existing project. To create a project based on an existing project, select *Create based upon an existing project*. This creates a new project with the same name and files as the selected project.

A new project is created in Capture and a blank schematic entry page with the name PAGE1 is displayed, as shown in Fig. 9.3.

Fig. 9.1 New project dialog box

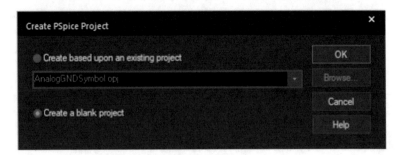

Fig. 9.2 Create PSpice project dialog box

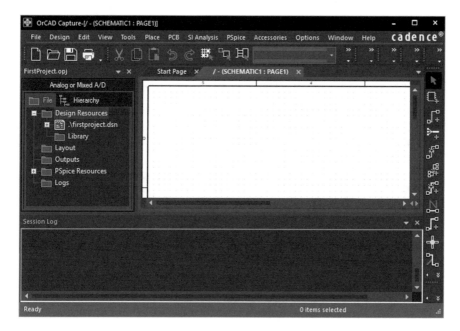

Fig. 9.3 Capture window with new schematic page

Next, you will create a simple schematic by placing one DC voltage source, one resistor, a capacitor and a ground. You will then connect the placed parts and change a few part properties. Figure 9.4 shows the circuit.

Fig. 9.4 Circuit design

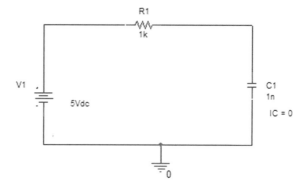

9.3   Placing and Connecting Parts

To place the parts that can be simulated in PSpice, do the following:

1. Choose the specific part from the *Place → PSpice Component* menu options.

   - To place a resistor, choose *Place → PSpice Component → Resistor.*
   - To place a capacitor, choose *Place → PSpice Component → Capacitor.*
   - To place a DC voltage source, choose *Place → PSpice Component → Source → Voltage Sources → DC.*
   - To place a Ground, choose *Place → PSpice Component → PSpice Ground.*

   Note: Right-click and choose *Rotate* to rotate a part.
   Click on the schematic canvas.
   Press Escape to end the placing of parts.
   To connect the placed parts, do the following:

1. Choose *Place → Wire*.

   Click the respective pins to connect them.
   Press Escape when done.

9.4   Searching PSpice Models or Parts

The PSpice® model library includes 34,000 + models for different devices, such as BJTs, JFETs, MOSFETs, IGBTs, SCRs, discrete, operational amplifiers, optocouplers, regulators and PWM controllers, from various semiconductor manufacturers. Each of these parts or models is categorized and listed with a brief description. Use the *Place → PSpice Component → Search* to open the PSpice search panel to browse, search and place parts.

9.5 Editing Part Properties

The placed parts have their own default properties and values. You can edit many of these properties.

1. Double-click *OVdc* for the voltage source to open the Display Properties window, as shown in Fig. 9.5, and edit the value to 5Vdc.

   You can also specify a *Display Format*, if needed.
   Select the capacitor, right-click, and choose *Edit Properties* to change property values.
   Click in the value column of the property *IC* and enter 0 to specify an initial condition of 0 V, as shown in Fig. 9.6.
   To display the property and value, click *Display* and then select *Name and Value*, as shown in Fig. 9.7.
   Click *OK* to close the window.
   Click *Apply* and close the Property Editor window.

Fig. 9.5 Display properties dialog box

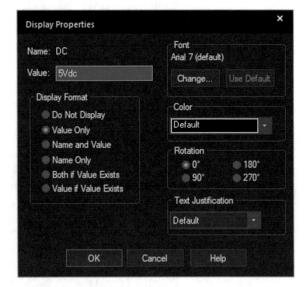

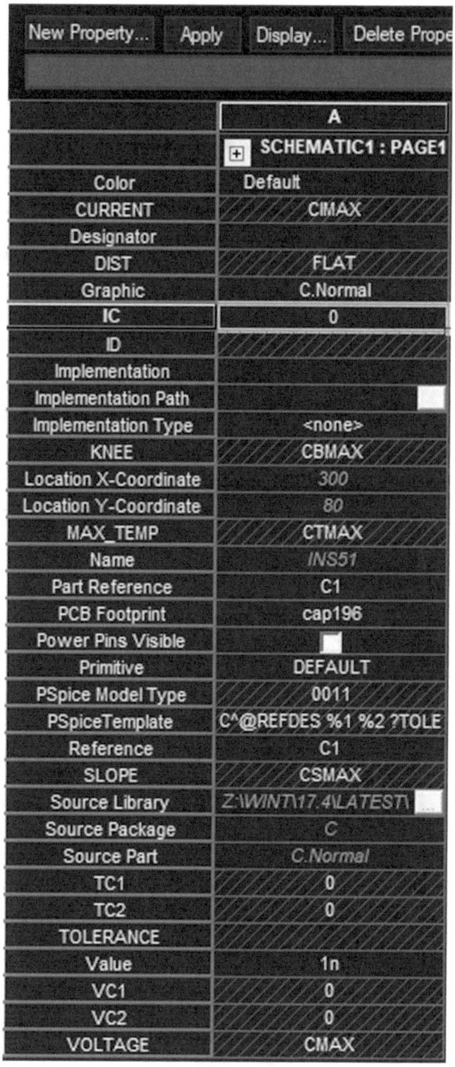

Fig. 9.6 Editing part properties

	A
⊞	SCHEMATIC1 : PAGE1
Color	Default
CURRENT	CIMAX
Designator	
DIST	FLAT
Graphic	C.Normal
IC	0
ID	
Implementation	
Implementation Path	
Implementation Type	<none>
KNEE	CBMAX
Location X-Coordinate	300
Location Y-Coordinate	80
MAX_TEMP	CTMAX
Name	INS51
Part Reference	C1
PCB Footprint	cap196
Power Pins Visible	
Primitive	DEFAULT
PSpice Model Type	0011
PSpiceTemplate	C^@REFDES %1 %2 ?TOLE
Reference	C1
SLOPE	CSMAX
Source Library	Z:\WINT\17.4\LATEST\
Source Package	C
Source Part	C.Normal
TC1	0
TC2	0
TOLERANCE	
Value	1n
VC1	0
VC2	0
VOLTAGE	CMAX

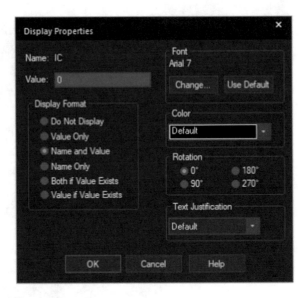

Fig. 9.7 Displaying property and value

9.6   Running a Simulation

To run a simulation, you must first create a simulation profile.

To create a simulation profile, do the following:

1. Choose *PSpice → New Simulation Profile* and then specify a name in the New Simulation dialog box as shown in Fig. 9.8.

Note: You can click the Open Design' button to open a project with a simulation profile created for you and then choose *PSpice → Edit Simulation Profile* to open the profile and view its settings.

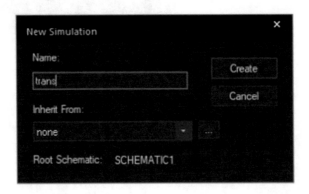

Fig. 9.8 New simulation dialog box

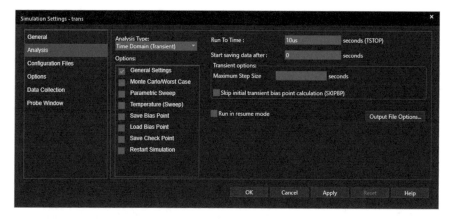

Fig. 9.9 Simulation settings dialog box

Click *Create*.

The Simulation Settings dialog box appears.

Change *Run to time* value to 10 μs to run the simulation for 10 μs, as shown in Fig. 9.9.

In this dialog box, you can specify the Analysis type, select different options and set various parameters for the simulation profile.

Click *OK* to accept the changes and save the profile.

Choose *PSpice → Run* to start the simulation.

The Probe window, as shown in Fig. 9.10, opens after simulation is complete.

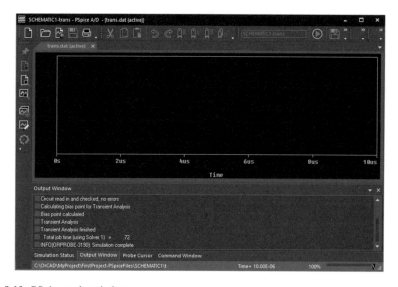

Fig. 9.10 PSpice probe window

Viewing Results

You will display the waveform for the voltage at *pin 2* of the capacitor, C_1.
To display the waveform, do the following:

1.   In PSpice, choose *Trace* → *Add Traces* to open the Add Traces window.
2.   Select $V(C_1:2)$, as shown in Fig. 9.11, and click *OK*.

The resultant waveform is displayed in the Probe window, as shown in Fig. 9.12.

Fig. 9.11   Add traces window

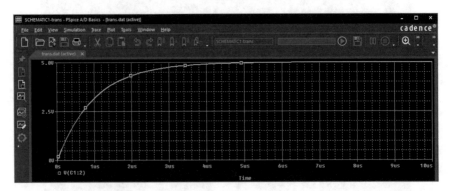

Fig. 9.12   Resultant waveform

9.7 Transient Analysis

Time-domain (transient) analysis plots the outputs as a function of time. For example, you might want to plot the voltage, $v_c(t)$, across a capacitor in a RC circuit over a period, say for 3 ms.

In a laboratory, measuring instruments, such as voltmeters and ammeters, are used to measure current and voltage. Oscilloscopes are then used to display the output as a trace. PSpice is a simulation tool that generates signals, emulating measuring instruments and displays the traces in the Probe window, emulating an oscilloscope.

In this chapter, you will:

- Run transient analysis for a switched series RC circuit to plot the voltage and current across a capacitor.
- Run transient analysis on a switched RLC circuit.
- Run transient analysis on a series RC circuit with an AC source to plot the voltage across a capacitor.

9.7.1 Transient Analysis of a Switched RC Circuit

In the circuit in Fig. 9.13, the switch U_1 is closed at 0 s and opens after 50 s. The switch U_2 is open at 0 s and closes after 50 s. The capacitor C_1, with initial voltage IC = 0 V, is charged for 50 s and then discharges through U_2. You will use PSpice to observe the current and the charging and discharging voltage across the capacitor.

Performing a time-domain (transient) analysis of the circuit for 100 s, you can observe the voltage across the capacitor, while charging for 50 s and then discharging for the remaining 50 s.

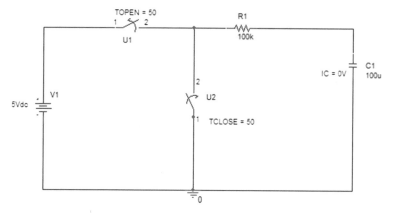

Fig. 9.13 Switched series RC circuit

1.   Start the simulation by choosing *PSpice → Run*.

Add a trace (*Trace → Add Trace*) in PSpice for $V(C_1{:}2)$, as shown in Fig. 9.14, to observe the voltage across the capacitor

Add a *Y* axis (*Plot → Add Y Axis*) and then add a trace for $I(R_1)$ to observe the current.

The result is shown in Fig. 9.15.

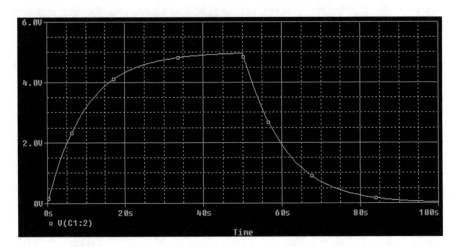

Fig. 9.14 Waveform for voltage across capacitor

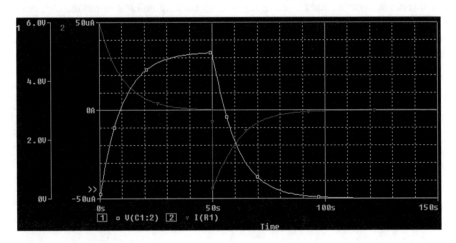

Fig. 9.15 Waveform for current

9.7.2 Transient Analysis of a Switched RLC Circuit

In the circuit of Fig. 9.16, the switch U_1 is open for 1 ms resulting in an open circuit for that time. The switch U_2 opens after 5 ms, and the switch U_3 closes after 5 ms.

From 0 s to 1 ms, the circuit is open. From 1 to 5 ms when both U_1 and U_2 are closed, the circuit is closed through capacitor C_2 and inductor L_1, and the capacitor C_2 is charged. At 5 ms, the switch U_2 opens, and U_3 closes. This leads to a discharge of C_2, charging C_1 in the process. Use PSpice to observe the charging and discharging of the capacitors. The charging and discharging of the capacitors can be observed by running a transient analysis for 10 ms and then displaying the traces for potential difference between the two pins of the capacitors as shown in Fig. 9.17.

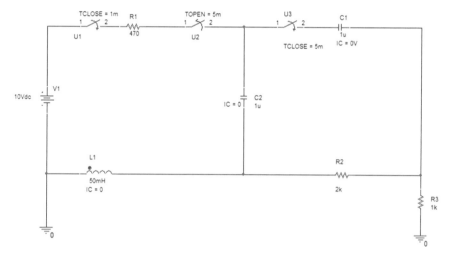

Fig. 9.16 Switched RLC circuit

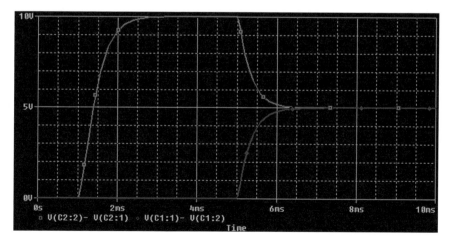

Fig. 9.17 Waveform for switched RLC circuit

Note: In PSpice, while adding traces, you can perform mathematical operations as in this example. For more information, refer *PSpice User Guide*.

9.7.3   Transient Analysis of a Circuit with an AC Voltage Source

In Fig. 9.18, an AC source, V_1, is connected in series to a resistor and a capacitor. The voltage source produces a sinusoidal signal of frequency 1 kHz and with amplitude of 1 V.

1.   Simulate the circuit for 4 ms with a maximum step size of 4 μs.
2.   Add a trace $V(V_1:+)$ to observe the voltage at the positive end of the voltage source.
3.   Add a trace $V(C_1:2)$ to observe the voltage across the capacitor as shown in Fig. 9.19.

   In a circuit with an AC source, the phase difference between the voltage and current across the capacitor is of interest. For the voltage, plot the trace $V(C_1:2)$.

   For the current, $I(C_1)$ shows the flow from top to bottom, but the reverse is of interest. Therefore, add a Y axis and plot the trace $-I(C_1)$. The resultant waveforms are shown in Fig. 9.20.

Note: In PSpice, while adding traces, you can perform arithmetic operators as in this example. For more information, refer *PSpice User Guide*.

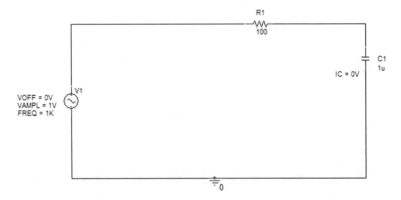

Fig. 9.18 RC circuit with AC source

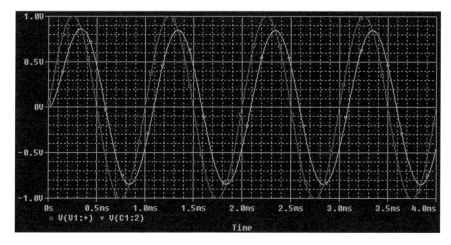

Fig. 9.19 Waveform with AC voltage source

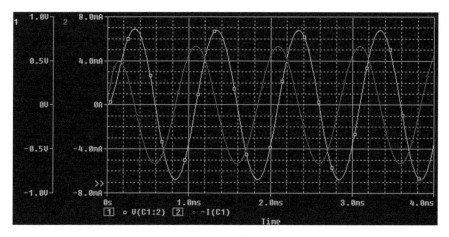

Fig. 9.20 Waveform for current and voltage phase difference

9.7.4 Transient Analysis of a RC Circuit with Switch

In Fig. 9.21, a 1 farad capacitor is connected in series with a switch and a 10 Ω resistor. The switch closes at 0 s and remains closed thereafter. The capacitor has an initial voltage of 5 V.

Plot the discharge of the capacitor and determine the voltage at the time constant of the circuit, which in this case is 1×10 or 10 s.

Fig. 9.21 Waveform for
current and voltage phase
difference

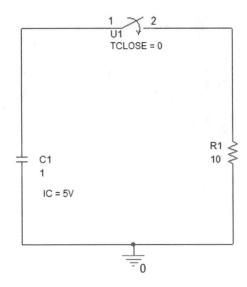

9.8   AC Analysis

AC sweep is a frequency response analysis. AC analysis is performed on linear circuits with a single sinusoidal source. Unlike transient analysis, where output is given as a function of time, the output of AC analysis is a phasor. A phasor represents both the amplitude and phase of the function without using time. You can perform AC analysis to determine AC gain for amplifiers or perform any network analysis to determine node voltage magnitude and phase. In AC analysis, PSpice calculates the small-signal response of a circuit to a combination of inputs by transforming it around the bias point and treating it as a linear circuit. In this chapter, you will:

- Perform steady-state AC analysis for a series RC circuit to determine the nodal voltage and the current across a capacitor. In a steady-state analysis, you find the phasor voltages and currents at a single frequency.
- Perform AC sweep analysis on a series RC circuit, that is, determine the frequency response of the circuit over a range of frequencies.
- Explore the decade and octave variations of the logarithmic scale along with PSpice functions, such as DB () and Bode Plot.

9.8.1   Steady-State AC Analysis of a RC Circuit

In the circuit in Fig. 9.22, which is a first-order low-pass filter, an AC voltage source is connected in series with a resistor and a capacitor. A linear circuit with an AC

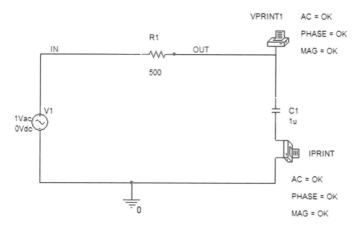

Fig. 9.22   Series RC circuit with AC voltage source

voltage source is required for AC analysis. You will perform steady-state analysis to determine the voltage at node *OUT* and the current across the capacitor.

Note: *VPRINT1* and *IPRINT* are used in the circuit to print or write simulation results to the output file. VPRINT1 is used to print or write the magnitude and phase of the voltage at the node it is connected. Similarly, IPRINT is connected in series with the capacitor to print the magnitude and phase of the current across the capacitor. In both VPRINT1 and IPRINT, the fields *AC*, *PHASE* and *MAG* must have a value, say *OK* or *YES*, to specify that the magnitude and phase are to be printed for AC.

AC analysis is performed for *Start Frequency* and *End Frequency* of 1 K and Total Points 1 as shown in Fig. 9.23.

To observe the output file, do the following:

1.   Choose View → *Output File* in PSpice.

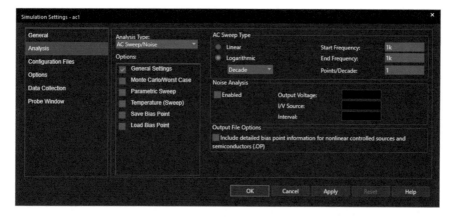

Fig. 9.23   Simulation settings for AC sweep—single frequency

```
****      AC ANALYSIS                        TEMPERATURE =    27.000 DEG C

******************************************************************************


  FREQ        VM(OUT)     VP(OUT)

   1.000E+03   3.033E-01  -7.234E+01
.
**** 06/19/14 15:20:58 ***** PSpice 16.6.0 (October 2012) ***** ID# 0 ********

 ** Profile: 'SCHEMATIC1-AC1'  [ D:\testcase\filters\ac_analysis-pspicefiles\schematic1\ac1.sim ]

   ****      AC ANALYSIS                        TEMPERATURE =    27.000 DEG C

******************************************************************************


  FREQ        IM(V_PRINT4) IP(V_PRINT4)

   1.000E+03   1.906E-03   1.766E+01


            JOB CONCLUDED
```

Fig. 9.24 Output file with current and voltage values

2. Browse to the section in the simulation output file that displays the phasors for
   the current and voltage as shown in Fig. 9.24.

   For voltage at the node OUT, the magnitude, VM(OUT), is 3.033E–01 V or
   0.303 V, and the phase, VP(OUT), is −7.234E+01° or −72.34°.
   For the current through the capacitor, the magnitude, *IM (V_PRINT4)*, is 1.906E-
   03A or 0.002A (approximately) and the phase, *IP (V_PRINT4)*, is 1.766E + 01° or
   17.66°.

Note: The output is for only one frequency (1.000E + 03 or 1 K) because this is
a steady-state analysis performed for that frequency value. Also, the voltage across
the capacitor lags the current by 90°.

9.8.2   AC Sweep of a RC Circuit

You will perform AC sweep analysis to determine the gain of the circuit shown in
Fig. 9.22.
   In the steady-state analysis, you run the simulation for a single frequency. For a
sweep analysis, you will specify a range of frequencies as shown in Fig. 9.25. Also
note that decade variation of the logarithmic scale is used for the sweep.
   The resultant waveform is shown in Fig. 9.26. This waveform uses the DB()
function available in PSpice to calculate and plot the gain of the circuit.
   To add the trace, do the following:

1. Open Add Traces window (*Trace → Add Traces*).
2. Click *DB()* under *Functions and Macros*.
3. Click *V(OUT)*.

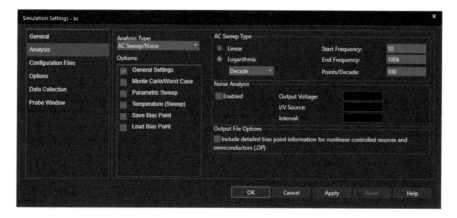

Fig. 9.25 Simulation for AC sweep analysis

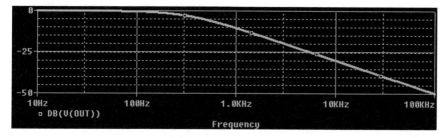

Fig. 9.26 Resultant waveform for AC analysis

9.8.3 Using Plot Window Templates to plot Gain and Phase Response

You will perform AC analysis of the circuit shown in Fig. 9.22 and use the Bode Plot function of the Plot Window Templates to determine the gain and phase.

1. Specify a range of frequencies and use an octave variation of the logarithmic scale as shown in Fig. 9.27.

To obtain the waveforms for gain and phase, use the Bode Plot function from under Plot Window Templates and specify V(OUT) as the parameter, as shown in Fig. 9.28.

The resultant waveforms are shown in Fig. 9.29.

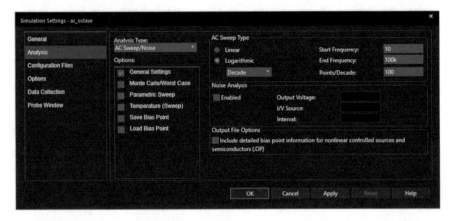

Fig. 9.27 Simulation for AC sweep analysis

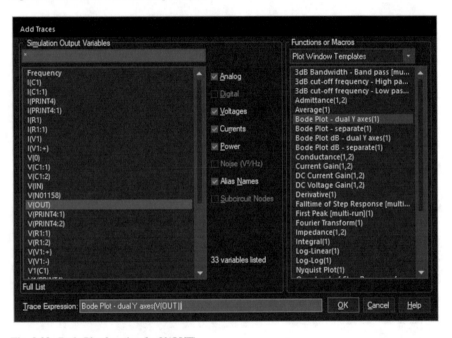

Fig. 9.28 Bode Plot function for V(OUT)

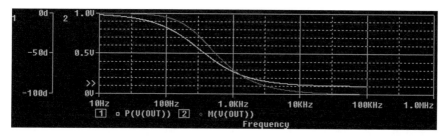

Fig. 9.29 Phase and gain waveforms for V(OUT)

9.8.4 Noise Analysis—Plotting Voltage Spectral Density

You will perform noise analysis of the circuit shown in Fig. 9.22 to determine the total voltage spectral density. In this circuit, the noise is generated by the resistor.

Note: The noise generated by a resistor can be modeled as a noiseless resistor in series with a noise voltage. The resistor noise is spread uniformly over frequency giving a flat spectral density, resulting in white noise.

1. Specify a range of frequencies and use a decade variation of the logarithmic scale as shown in Fig. 9.30.

Note: Noise Analysis is enabled with $V(OUT)$ as the output voltage and V_1 as the input voltage. The noise is to be analyzed at an interval of 1 kHz as specified.

To obtain the total voltage spectral density, add the trace NTOT(R_1) from the Add Traces window.

The resultant waveform is shown in Fig. 9.31.

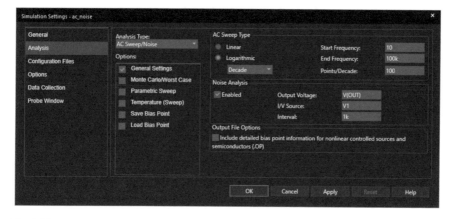

Fig. 9.30 Simulation for AC sweep analysis

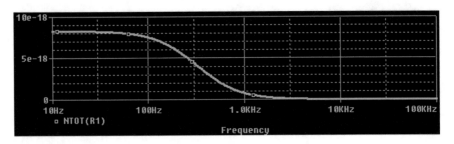

Fig. 9.31   Voltage spectral density for resistor R_1

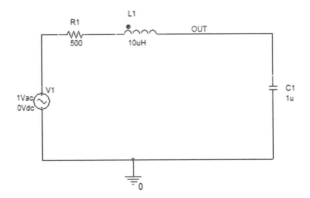

Fig. 9.32   Second-order low-pass filter

9.8.5   Second-Order Low-Pass Filter Gain Simulation

Figure 9.32 shows a second-order low-pass filter. Use PSpice to plot the gain of the filter.

9.9   Developing a Model of Real-Life Inductor

The circuit shown in Fig. 9.33 models a real-life inductor and analyzes the effect of temperature on an inductor and how it can impact the rest of the circuit.

Consider a typical inductor with following specification:

Inductance value at 27 °C	1 μH
DCR	0.14 Ω
SRF	340 MHz
Linear temperature coefficient	0.014 per °C
Quadratic temperature coefficient	0.0001 per °C

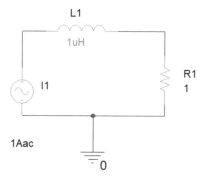

Fig. 9.33 Inductor test circuit

Use the PSpice Modeling Application to create a SPICE model during schematic entry. To create a model for the inductor, do the following:

1. Choose *Place → PSpice Component → Modeling Application.*

   Select *Inductor* under *Passives* to open Inductor Modeling application.
   Specify the electrical configuration for the inductor, as shown in Fig. 9.34.
   Click *Place* to place an inductor symbol and associate a SPICE model as per electrical specification provided.
   Use a very simple circuit, as shown in Fig. 9.1, to verify the inductor electrical characteristics by using frequency-domain analysis.

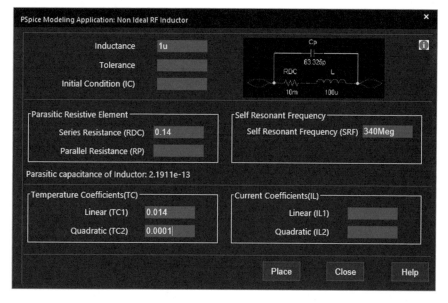

Fig. 9.34 PSpice modeling application—inductor

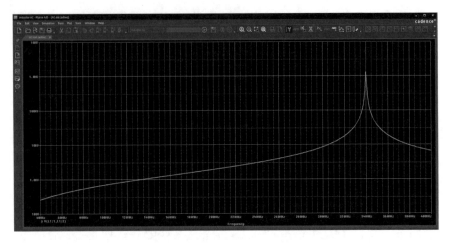

Fig. 9.35 Inductor impedance variation with frequency

Perform AC sweep analysis for a frequency range of 40 MHz to 400 MHz, and plot the voltage across the inductor.

Since the input source is 1Amp sinusoidal AC current source, voltage across inductor also represents the impedance of inductor. Plot of this voltage is shown in Fig. 9.35, and this represents impedance variation of inductor with respect to frequency. The peak at 340 MHz shows the resonance, and this matches with input specification of inductor.

You can simulate other test fixtures to verify the other electrical characteristics.

This enables you to quickly create and simulate real-life inductor models with or without parasitic. PSpice also offers several other modeling applications.

9.10 Temperature Sweep of a RL Circuit and Analyzing the Effect of Temperature on Circuit

You have defined the temperature coefficient of the inductor. To see if circuit response changes because of temperature change, set up the simulation profile to simulate at different temperatures (temperature sweep). Circuit should have different resonant frequency as inductance value changes due to change in temperature, whereas other parasitic may not change or change in equal amount.

To set temperature sweep, do the following:

1. Create a new simulation profile of type AC analysis (frequency sweep).
    You can also modify an existing profile to add temperature sweep.
2. To add temperature sweep, select Temperature (Sweep).
3. Select *Repeat the simulation for each of the temperatures* and enter a list of temperatures separated by space in field (10 27 50 85), as shown in Fig. 9.36.

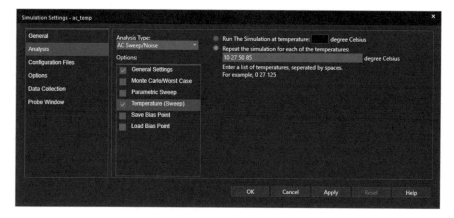

Fig. 9.36 Simulation setup for temperature sweep

On simulation run PSpice first calculates the inductance value for the current simulation temperature and generates frequency response of circuit using the calculated
inductance value. Simulation result shows four plots for inductance impedance, as
in Fig. 9.37.

The first curve in green plots the frequency response for the first temperature value
in the sweep list, which is 10 °C. Remaining three curves are for the other sweep
temperatures. The red plot represents the circuit response at 27 °C (the nominal
value) and shows the resonance frequency, which is 340 MHz.

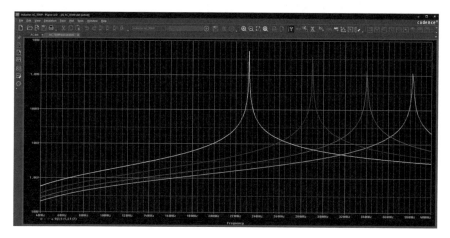

Fig. 9.37 Inductor impedance variation with frequency for different temperatures

9.11   PSpice for Digital and Mixed-Signal Analysis

Digital simulation is the analysis of logic and timing behavior of digital devices over time. Mixed-signal analysis is the analysis of a circuit that includes both digital and analog devices. Most of the real-life circuits are mixed-signal circuits. PSpice includes a dedicated event-driven kernel for simulation of digital devices. This capability is not available in any other SPICE simulator, most of SPICE simulator mimics the digital devices behavior using SPICE behavioral models, lacking true digital or mixed-mode simulation.

PSpice digital device model has the following three main characteristics:

- Functional behavior: Gate-level and behavioral digital primitives comprising the sub-circuit. This controls the behavior of a digital device.
- I/O behavior: I/O model, interface sub-circuits and power supplies related to a logic family. This specifies information specific to the device's input/output characteristics or the relationship between drive resistances and output strengths.
- Timing behavior: One or more timing models, pin-to-pin delay primitives, or constraint checker primitives. This specifies propagation delays and timing constraints, such as setup and hold times. Within a timing model, there may be one or more types of parameters, such as propagation delay and setup times and hold times. Further, each parameter is divided into three values based on device datasheet: minimum (MN), typical (TY) and maximum (MX).

Using above mechanism, PSpice offers a very methodical approach to design and develop digital device models and use the models for different applications easily. For example, you can develop an AND gate functional model and associate it with different TIMING and I/O behavior models to realize different AND gate modes for different logic families, such as CMOS and TTL-ECL. The advantage of having two models is that, while timing information is specific to a device, the input/output characteristics are specific to a whole logic family. Many devices in the same family reference the same I/O model, but each device has its own timing model.

PSpice performs detailed timing analysis subject to the constraints specified for the devices. For example, flip-flops perform setup checks on the incoming clock and data signals. PSpice reports any timing violations or hazards as messages written to the simulation output file and the waveform data file.

9.11.1   Simulating Analog-to-Digital Converter (ADC)

Analog-to-digital converter (ADC) converts an analog input voltage to its corresponding digital value. While the analog input can take an infinite number of values, as it is a continuous signal, the output can be selected only from a finite set of digital values depending on the converter's resolution. Thus, an ADC approximates each of the input levels with one of the digital codes. This is generally done by using a set of reference voltages corresponding to each code and comparing the analog input value with the reference voltages until the best approximation is found for the analog input voltage. The analog input is approximated with the nearest smaller reference level. For an m-bit ADC, the digital output D (evaluated in decimal representation) is calculated using the following equation:

$$D = \left[(2^m \cdot A)/V_{ref}\right]$$

Figure 9.38 shows an 8-bit ADC circuit implementation in PSpice. Since analog input is a continuously varying signal, it must be held at a value level for analog-to-digital conversion to take place for that level. This is achieved through the sample and hold architecture. Here, V_5 is the pulse controlling the sample and hold circuit. That is, when V_5 is high, input is sampled as the voltage-controlled switch S_3 is closed and the capacitor starts charging to voltage V_7 (input voltage), which is the analog ramp signal. When V_5 goes low, switch S_3 is open and voltage across capacitor is held for digital conversion. Another voltage-controlled switch S_4 is open during the sampling phase and closes during the hold phase, so that the analog input is passed to the input pin only when it is held at some DC or constant voltage level. A source

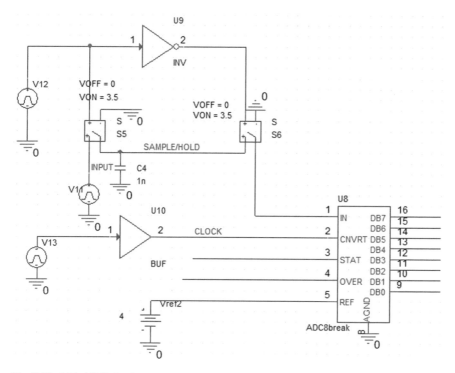

Fig. 9.38 8-bit ADC circuit

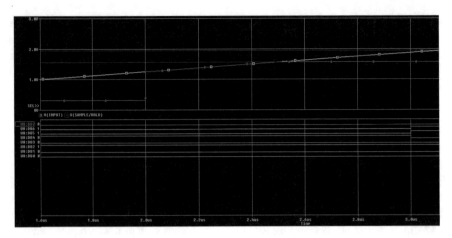

Fig. 9.39 8-bit ADC circuit simulation results

pulse V_6 is the clock input. Data is converted to digital form at every rising edge of the clock and stays for one clock period.

For an analog input of 1.5634 V at some instant of time in the ramp input, the output code (in decimal representation) can be calculated as:

$$D = [(2^8 * 1.5634)/4] = [100.0576] = 100$$

The binary representation of 100 is 01100100. Figure 9.39 shows the simulation results for the analog ramp input voltage held at above-mentioned voltage levels by the *sample and hold* circuitry (shown here in red), clock pulse and corresponding 8-bit digital output. Here, input voltage is held at level 1.563 V, and ADC is converting this into digital value. The digital output bits DB 7 to 0, 8 bits of DB show the corresponding digital output.

You can turn on cursor window in PSpice to measure the exact voltage levels for these individual bits and you will notice that DB[7-0] 01100100 matches the calculated value, as shown in Fig. 9.40.

You can also plot the waveforms as a BUS notation instead of individual bits, making analysis easier.

In the following figure, the topmost waveform named *ADC* plots DB[7:0] as BUS. You can also select the different numbering system to display value. PSpice supports binary (used below), hexadecimal, octal and decimal number systems.

Syntax for this is {*bus_prefix*[*msb:lsb*]}[;[*display_name*][;*radix*]]; for example, {U8:DB[7:0]};ADC;B. Here, the last B specifies the binary number system—D plots in decimal, H in hexadecimal and O in the octal number system, as shown in Fig. 9.41.

Trace Color	Trace Name	Y1	Y2
	X Values	3.1073u	3.1073u
	U8:DB7	0	0
	U8:DB6	1	1
	U8:DB5	1	1
	U8:DB4	0	0
	U8:DB3	0	0
	U8:DB2	1	1
	U8:DB1	0	0
	U8:DB0	0	0
CURSOR 2	V(INPUT)	1.9421	1.9421
CURSOR 1	V(SAMPLE/HOLD)	1.5634	1.5634

Fig. 9.40 Exact voltage levels for the individual bits

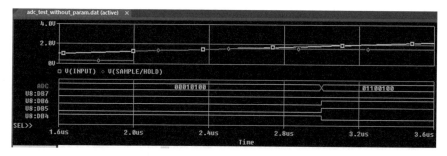

Fig. 9.41 Bus notation

9.11.2 Simulating Worst-Case Timing Simulation

Digital worst-case timing simulation is useful when timing of signals is critical for the proper operation of a design. In a simple timing simulation (using one of MIN, TYP or MAX delay), signal propagation through digital devices is normally represented as an *instantaneous* transition. During the worst-case timing simulation, the effects of individual component delay ranges are propagated throughout the circuit. The transitions take both the MIN as well as the MAX delay characteristics of their propagation paths. Transitions may be thought of as regions of signal ambiguity due to the uncertainty of which delay value (MIN, MAX or somewhere in between) applies to each component used in the design. PSpice represents this type of signal ambiguity with Rising (R) and Falling (F) logic levels.

The circuit shown in Fig. 9.42 has been configured to simulate design for maximum delay and worst-case delay.

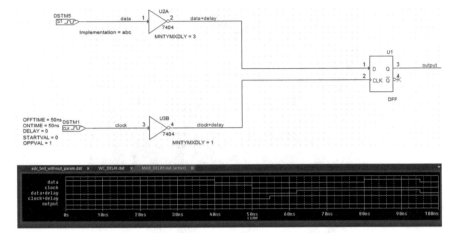

Fig. 9.42 Configuration to simulate design for maximum delay and worst-case delay

Notice the propagation delay for data and clock signals in the above simulation result. Realizing that any two (or more) instances of a particular type of component may have propagation delay values anywhere within the published range, designers are faced with the problem of ensuring that their products are fully functional when they are built with combinations of components having delay specifications that fall (perhaps randomly) anywhere within this range. Simulating a design with typical or maximum delay may not uncover all possible violations as demonstrated by example. True digital worst-case simulation, as provided by PSpice, does just that. If you run the same circuit in worst-case timing analysis mode, PSpice will highlight all possible timing violations, such as timing mismatch between data and clock. Following waveform, Fig. 9.43 shows one timing violation condition on the above circuit. Waveform portion marked in yellow or gray indicates ambiguous or invalid signals due to timing issue.

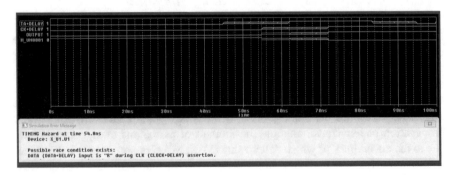

Fig. 9.43 Timing violation condition

9.12 PSpice Advance Analysis

For any real-life design, functional simulation in ideal or near ideal conditions is not sufficient. You need to consider a lot of other factors which are beyond functional simulation. For example, a design that needs to be mass produced must factor in component tolerances so that the circuit functions properly throughout the range of all the possible combinations of component values. PSpice Advanced Analysis provides a set of advanced simulation engines and post-processing modules that are used in conjunction with PSpice simulation to maximize design performance, yield, cost-effectiveness and reliability. These advanced simulation capabilities— Sensitivity, Monte Carlo, Smoke (Stress), Optimizer and Parametric Plotter—help you deal with the problems of manufacturing variations of electronic components by providing you with an in-depth understanding of circuit performance that goes beyond basic validation.

9.12.1 Smoke (Electrical and Thermal Stress) Analysis

Smoke analysis performs electrical stress analysis of circuit components under different simulation conditions. This helps identify component stress and potential failures due to power dissipation, increases in-junction temperature, secondary breakdowns or violations of voltage and current limits. Smoke analysis compares circuit simulation results with the component's specification and applies configured deration or margins in safe operating limits, and if the limits are exceeded, identifies the problem parameters, as shown in Fig. 9.44.

Fig. 9.44 Smoke analysis

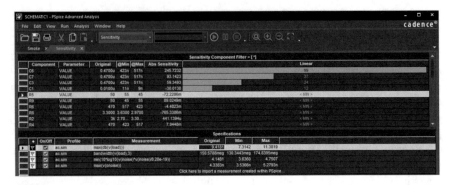

Fig. 9.45 Sensitivity analysis

9.12.2   Optimizing a Circuit

This is a circuit optimization engine. You can set multiple goals such as bandwidth or power loss of a device and let optimizer find the circuit component values to meet that goal. You can also set goals in terms of waveform; for example, set measurements from a digital storage oscilloscope as the goal for a given circuit. Optimizer will find circuit component value to meet that specific waveform.

9.12.3   Sensitivity Analysis

Sensitivity analysis identifies the component parameters that are critical to your circuit performance goals. It examines how a component's inherent manufacturing variations affect circuit behavior both individually and in comparison with other components by varying manufacturing tolerances to create worst-case (minimum and maximum) results. You can set more than one goal and set tolerances for any number of components for sensitivity analysis in Advanced Analysis, as shown in Fig. 9.45.

9.12.4   Monte Carlo Analysis

Monte Carlo analysis predicts the behavior of a circuit statistically when multiple components are varied within their tolerance ranges. By changing all the parts of your circuit randomly within their tolerances over a number of simulations, you can approximate the yield of building numerous boards. Use Monte Carlo analysis to set multiple goals across different types of analysis and to run all the analyses together in one go, thus saving significant simulation time, as shown in Fig. 9.46.

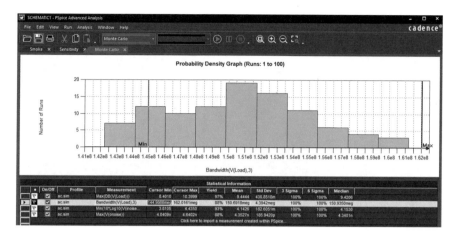

Fig. 9.46 Monte Carlo analysis

9.12.5 Parametric Plotter

Parametric Plotter enables sweeping of multiple parameters for a given circuit. It also provides an efficient way for you to analyze sweep results, sweep any number of design and model parameters (in any combination) and view the results in Parametric Plotter or Probe in a tabular or plot form.

9.13 Getting Access to PSpice

9.13.1 PSpice Trial

Cadence design system offers you a free, fully functional, limited period PSpice.

Register yourself at https://www.orcad.com/pspice-free-trial to get access to a full-fledged version of latest Cadence® PSpice® Simulation software for free including OrCAD Capture, PSpice A/D, more than 34,000 models, PSpice Advanced Analysis and the PSpice MATLAB® interface.

9.13.2 PSpice for Academic

PSpice Academic Program enables students and educators access the latest version of PSpice software available for students and educators. This program offers you fully functional, professional version of PSpice software used by the world-wide design community every day at a very nominal cost. Visit https://www.pspice.com/academic-user for more details and to register yourself for access.

Bibliography

1. AN574: Printed Circuit Board (PCB) Power Delivery Network (PDN) Design Methodology, ALTERA, May 2009.
2. Application Note: Transient Voltage Suppressors (TVS Diode) Applications Overview, Littlefuse, 2015.
3. Basics of Ideal Diodes, Application Report SLVAE57–May 2019. USA: Texas Instruments.
4. Carr, J. (2001). *Antenna toolkit* (2nd ed.). Newnes.
5. Carr, J. J. (2002). *RF components and circuits* (1st ed.). Newnes.
6. Carter, B. (2000, November). *A single-supply op-amp circuit collection, application report SLOA058.* USA: Texas Instruments.
7. Carter, B. *Understanding op amp parameters* (Chap. 11). Literature Number SLOA083, Excerpted from Op Amps for Everyone Literature Number: SLOD006A. USA: Texas Instruments.
8. Cockrill, C. (2011, September). *Understanding schmitt triggers.* Application Report SCEA046. USA: Texas Instruments.
9. Cordell, B. (2011). *Designing audio power amplifiers.* New York: McGraw-Hill.
10. Desoer, C. A., & Kuh, E. S. *Basic circuit theory.* New York: McGraw-Hill.
11. Gray, P. E., & Meyer, R. E. *Analysis and design of analog integrated circuits.* New York: Wiley.
12. Gray, P. E., & Searle, C. L. (1971). *Electronic principles.* New York: Wiley inc.
13. Green, D. (1986). *Modern logic design* (Chap. 1). UK: Addison-Wesley.
14. Groner, S. (2011, September). A new audio amplifier topology with push-pull transimpedance stage part 1. *Linear Audio, 2.*
15. http://www.johnhearfield.com/Eng/Schmitt.htm. Schmitt trigger design, 2010.
16. https://www.allaboutcircuits.com/technical-articles/how-to-design-schmitt-trigger-oscillators/, May 2018.
17. Jones, M. H. (1995). *A practical introduction to electronic circuits* (3rd ed.). Cambridge: Cambridge University Press.
18. Karki, J. (1998, April). Understanding Operational Amplifier Specifications, WHITE PAPER: SLOA011. Mixed Signal and Analog Operational Amplifiers, Digital Signal Processing Solutions. USA: Texas Instruments.
19. Kikkert, C. J. (2013). *RF electronics—Design and simulation* (Chap. 6). Townsville, QLD, Australia: James Cook University.
20. Kumar, P. M. (2014, April). *Crystal oscillator fundamentals and operation—Part II.* EDN Network. April 08, 2014.
21. Lacaita, A., & Sampietro, M. (2011). Circuiti Elettronici CittaStudiEdizioni.
22. LM107, LM118, AN-A. The Monolithic Operational Amplifier: A Tutorial Study, Literature Number: SNOA737. USA: Texas Instruments.
23. LM317 3-Terminal Adjustable Regulator, Texas Instruments, SLVS044X SEPTEMBER 1997–REVISED SEPTEMBER 2016.

© Springer Nature Switzerland AG 2020
R. Gastaldi and G. Campardo, *Electronic Experiences in a Virtual Lab*,
https://doi.org/10.1007/978-3-030-45179-0

24. LM555 Single Timer, 2002 Fairchild Semiconductor Corporation, January 2013.
25. LORD KELVIN'S SENSING METHOD LIVES ON IN THE MEASUREMENT ACCU-RACY OF ULTRA-PRECISION CURRENT-SHUNT MONITORS/CURRENT-SENSE AMPLIFIERS, APPLICATION NOTE 5761, © 2014 Maxim Integrated Products, Inc.
26. Mancini, R. (Ed.). (2002, August). *Op amp for everyone*. Design Reference, August 2002. Advanced Analog Products, SLOD006B. USA: Texas Instruments.
27. Mancini, R. (Ed.). (2002, August). *Op amps for everyone*. Design Reference. USA: Texas Instruments.
28. MC34063A, MC33063A, NCV33063A 1.5 A, Step-Up/Down/ Inverting Switching Regulators, Dec. 2003, Rev11, On Semiconductor.
29. Millman, J., & Halkias, C. (1972). *Integrated electronics* (Chap. 14). New York: McGraw-Hill.
30. Millman, J., & Halkias, C. (1973). *Integrated electronics, cap.17*. New York: McGraw-Hill.
31. Morris, R. L., & Miller, J. R. (Eds.). (1997). *Designing with TTL IC's*. New York: McGraw-Hill.
32. Nicoletti, R. (2013). *Choosing the right audio amplifier topology*. EEtimes-Asia.
33. NTC Thermistors Introduction to NTCs, Philips Components March 31, 1995.
34. Op Amp Circuit Collection, National Semiconductor, Application Note 31, September 2002.
35. Op Amp Common-Mode Rejection Ratio (CMRR), MT-042 TUTORIAL, Analog Device, 2009.
36. Overview of Two-Wire and Four-Wire (Kelvin) Resistance Measurements, Application Note.
37. Power integrity is more than decoupling capacitors…The Power Integrity Ecosystem, Keysight HSD Seminar Mastering SI & PI Design, KEYSIGHT technology.
38. Rectifiers Application Note: Physical Explanation, VISHAY GENERAL SEMICONDUC-TOR, Revision 16-Aug-11.3, Document Number: 84064.
39. Series, Number 3176, KEITHLEY A Tektronix Company, 2012.
40. Silicon Labs AN0016.2 Oscillator design considerations. http://www.silabs.com, July 2019.
41. The monolithic operational amplifier: A tutorial study, national semiconductor, Appendix A A, December 1974 Invited Paper—*IEEE Journal of Solid-State Circuits, SC-9*(6).
42. Transformerless Power Supplies: Resistive and Capacitive, MICROCHIP, AN954, 2004.
43. Transient Voltage Suppression diode application notes, Technical Note 4048, December 2017, Supersedes 2009, BUSSMANN series.
44. Trump, B. (2017). *The signal, a compendium of blog posts on op amp design topics*. USA: Texas Instruments.
45. Understanding Boost Power Stage in Switchmode Power Supplies, Texas Instruments Application Notes, March 1999.
46. Using Current Sense Resistors for Accurate Current Measurement, 06/18 e/N1812, BOURNS.
47. Using Decoupling Capacitors, Cypress, AN1032.
48. Varactor Diode or Varicap Diode, electronicsnotes. https://www.electronics-notes.com/articles/electronic_components/diode/varactor-varicap-diode.php.
49. Wells, C., Kay, A., Williams, I., & Green, T. CMRR TIPL 1231 TI Precision Labs—Op amps. USA: Texas Instruments.
50. Williams, J. (1985). *A designer's guide to: Innovative linear circuits as featured in EDN magazine*. Cahners Publishing Company.
51. www.st.com. AN2876—Oscillator design guide for STM8AF/AL/S, STM32 MCUs and MPUs, rev.12, January 2020.
52. Zumbahlen, H. (Ed.). (2008). *Linear circuit design handbook*. Analog Device, Inc., Elsevier.
53. μA741 General-Purpose Operational Amplifiers SLOS094G–NOVEMBER 1970–REVISED JANUARY 2018. USA: Texas Instruments.

Index

© Springer Nature Switzerland AG 2020
R. Gastaldi and G. Campardo, *Electronic Experiences in a Virtual Lab*,
https://doi.org/10.1007/978-3-030-45179-0

Printed in the United States
by Baker & Taylor Publisher Services